自见

The Insight Brings Us to the Future

T10 设计作品集（2020）

Collection of T10 Design (2020)

T10 设计　著

華中科技大學出版社
http://www.hustp.com
中国 · 武汉

前言
Preface

和光同尘，观往而知来

当大家看到这篇文字的时候，本书已出版，想至此，难掩内心激动。

一切缘起于 2014 年底，当时，我们同时获得了“Top10（宁波）优势力设计团队”奖，因着对设计和美学的共同热爱，走到一起，组成一个团队，希望在创作之余，能共同探究生活、美学、艺术乃至消费的哲学。

不知不觉，6 年过去了，每每翻看历次合影，变化之微妙尽在不言中。

6 年里，T10 邀请国内优秀设计师来宁波作交流；与高校联合，举办公益竞赛，激发学生的原创热情；一起游历海内外，拜访名师，拓展视野；与志趣相投的材料商合作，以产业链优势，做更多有社会价值和意义的事。

6 年里，聚散离合，起起伏伏，T10 成员在不断的磨合中日趋默契，亦更成熟。

当大家在各自的领域和项目中埋头奋进之时，疫情，却始料未及地肆虐起来。2020 年，真是特殊的一年。它几乎改变了所有人的计划，扯住了每个人急切前进的步伐。

在停顿的瞬间，我们开始回顾过往，反思自己。这便是“自见 The Insight Brings Us to the Future”这个书名的由来。

这本书是过去 6 年中，T10 成员在各自擅长领域内的部分作品的合集，包含文化、办公、商业和居住空间，是 T10 第一次以作品联合发声。

作品照见了我们不同的个性，但融合在一起，又充满共性。它让我们看见自己成长的空间和今后的路。作品，是我们力量的根本。

T10，正是倚赖设计之本，行有益之事，做一个虔诚的美学传播者。

自见，正如古老的道家智慧所启示的：和光同尘，观往而知来。

一方面放下自我，求同存异，以宽容谦受之量，融入尘世，发挥自己的积极能量；一方面，适时停下，审视过去，洞察自我，以期遇见更满意的未来。

以上，与诸位同仁和读者共勉。

T10 设计

2020 年，孟冬

Veil the Sharpness, Perceive Tomorrow with Yesterday

If you are reading this, then the book must have been published – we cannot tell you how excited we are to picture its publishing.

Everything started at the end of year 2014, when we were awarded "Ningbo Top 10 Excellent Design Teams", and then we formed a new team because we shared the same enthusiasm of design and aesthetics. Our original drive was to study the philosophy of life, aesthetics, arts, and even philosophy of consumption alongside designing.

In a blink of an eye, 6 years flew by. Whenever we see our old photos, we can sense everyone's change – the emotion is beyond words.

In the past 6 years, we invited outstanding architects for lectures; we encouraged creative ideas from college students by hosting contests together with universities; we visited numerous great architecture and architects to pursue a better practice in architecture; we found great partners to work with, and conducted business with advantages in industry chain to create more values towards the industry and society.

In the past 6 years, with encounters and departures, with ups and downs, we have been getting stronger. Teamwork is appearing and team members are growing.

Then suddenly, when we were all accelerating to accomplish our goals, the pandemic of COVID-19 burst out uncontrollably. 2020, a truly special year. Almost everyone was forced to stop and then create plan B. At the moment when we paused, we began to retrospect. This is where the book title "*The Insight Brings Us to the Future*" comes from.

This book contains parts of the most successful designing practices from each member in our team within the past 6 years, including culture, office, business and residence. This is also the first time we publish a book of collection works of the whole team.

All the pieces reflect our personalities and differences. On the other hand, we are full of similarities when we are infused with each other. The pieces let us see ourselves and foresee our paths. They are the roots of our power.

"T10+ Design Alliance" exists because we all love designing. Basing on that, we want to create more positive energy, and spread pure aesthetics.

To keep our initial goal everlasting, we follow the wisdom from Tao – veil the sharpness, perceive tomorrow with yesterday. We shall veil our ego to accept others, to sense the world with great tolerance, and to create positive energy. We shall also pause the journey to review the past, to examine ourselves, and thus to encounter a better future.

May you readers and we achieve this together.

T10 Design

Winter, 2020

目录
Contents

文化休闲 | Cultural Space

办公空间 | Office Space

商业展示 | Shop & Exhibition Space

居住空间 | Dwelling Space

文化休闲
Cultural Space

他要像一棵树栽在溪水旁，

按时候结果子，

叶子也不枯干。

——《圣经》和合本诗篇 1：3

He is like a tree planted by streams of water

that yields its fruit in its season,

and its leaf does not wither.

– Psalms 1: 3 NIV

左页：从门厅望向外面。4~5页：教堂内景。

Opposite: View outside the door. pp.4-5: Interior view of the church.

设计中的遇见——望春堂

Meeting in design – Wangchun Church

望春堂位于公园里，南面有一条静静的河，河边树木枝繁叶茂。设计之初，我在建筑与自然空间里寻找教堂应有的特性，感受神所要赋予它的建筑语言。

教堂的大门尺度不大，但可以将门外的绿树与河流框入我们的眼帘，户外的自然与室内的光辉由此共融。

At the beginning of the project, I looked for characteristics of the church through its architecture and the natural space. Located in a park, Wangchun Church has a river to its south which flows quietly past an abundance of trees. The door of the church is relatively small, but invites the shadow of the trees and the flowing river inside, blending the outdoor scenery with the indoor space.

左页：教堂入口。本页，上：窗边十字架；下：二层内景。

Opposite: The entrance of the Church. This page, top: Close view of the Cross by the window; bottom: View of the upper floor.

坐在教堂的条凳上，看着静谧的光洒落在凳子上，静静地感受神的存在。

Sitting on the bench in the church, as the calm light falls, one feels the god to be with you.

上：顶窗；下：楼梯。
Top: One of the top windows;
bottom: Staircase.

抬头仰望，光透过顶上的 6 个窗洒落在水泥地上。就如《圣经》上所记载：“我是世界的光。跟从我的，就不在黑暗里走，必要得着生命的光。”

The sunlight passes through six windows on the roof and falls on the concrete floor. Just as the *Bible* says: I am the light of the world. Those who follow me will not walk in darkness, but will find the light of life.

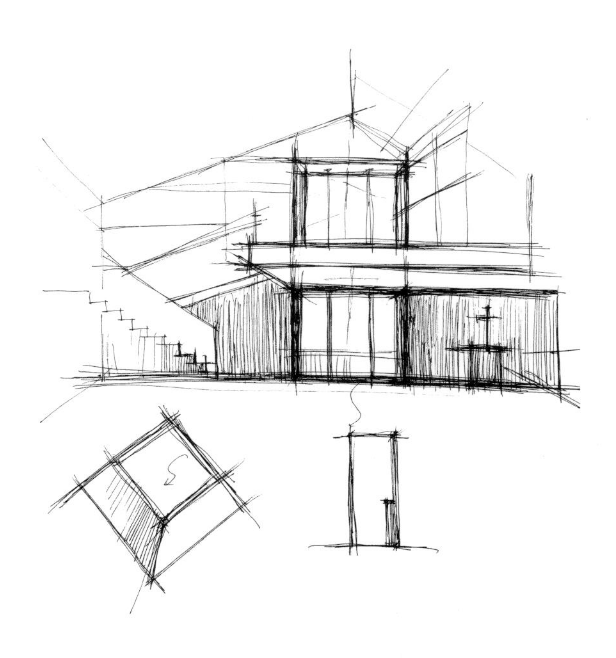

手稿，关于横剖面、门和顶窗。

Sketch of the cross section, door and top window.

教堂是神圣而宁静的，让浮躁的心灵在此得到平静，让世俗的灵魂得到洗涤，让惘然的生命重新看到希望。所以设计也必然是本质而虔诚的，用纯粹的内心、本质的材料和自然的光辉去呈现这个场所的精神。

A church is sacred and quiet in which the soul can rest in peace and lost people can find hope again. Therefore, the design of church must be pious as well. By means of a pure heart, essential materials and natural light, a church's spirit thus can be manifested.

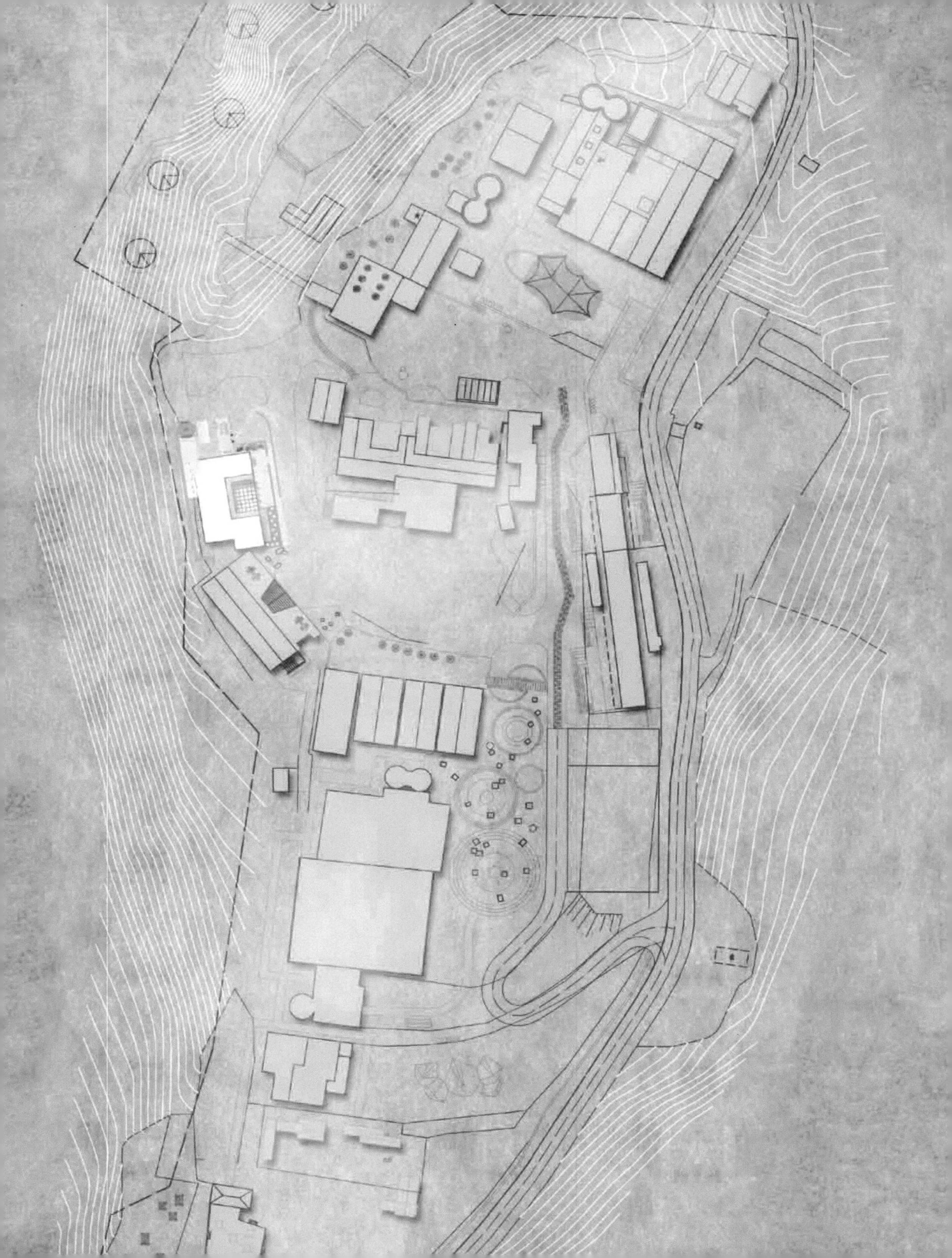

左页：申山乡宿总体规划图，白色为一号别院。12~13 页：黄昏时分的一号别院。

Opposite: The master plan of Shenshan rural Inn, with #1 villa highlighted. pp.12-13: View of the #1 villa at dusk.

秘境花园——申山乡宿一号别院

Secret garden – #1 Villa of Shenshan Rural Inn

项目位于衢州常山，定位为以度假和休闲为主的精品民宿。原本是一座 20 世纪 90 年代的三层砖混办公楼，平面呈“凹”字形，南北长 11 米，宽约 20 米，开间深度 5 米。

The project is located in Changshan, Quzhou city, and defined to be a botique B&B. It was originally a three-story brick-concrete office building in the 1990s. The plan is concave, 11 meters long, 20 meters wide and 5 meters deep.

下沉式休息区。
The sunken resting area.

一楼作为公共区，设计了下沉式休息区、厨房、餐厅、茶室和办公区，周边以景观连廊连接，并以景观水池和微缩景观的构建完成内部装饰，以自然原味的乡宿氛围，平衡周边的生态和景观，自身成景。

公共区近景。
Close view of the public area.

The first floor serves as a public area, with a sunken resting area, kitchen, dining room, tea room and office area. The surroundings are connected by a corridor. The pools and miniature landscape are utilized to decorate the interiors. The building gives some kind of natural rural atmosphere, which balances the surrounding ecology, and forms a scene by itself.

中庭。
The courtyard.

客房走廊。
View of the corridor connecting guestrooms.

整体建筑最大的改造在于凹口部分的围合和景观阳台的构建，不但利用了中部的空间，而且给予二楼、三楼的客房部分更多的景观和休闲空间。

The biggest renovation of the building lies in the enclosure of the notch and the construction of the landscape balcony, which not only takes good use of the space in the middle, but also gives the guestrooms more views and leisure space.

本页：客房陈设。右页：客房卫生间质朴的肌理。
This page: Furnishings in guestroom. Opposite: View of the bathroom, with simple and primitive texture.

一方面，设计师需要保留原建筑的主体结构和特色，对建筑肌理与形态进行改造，使之成为区域标志；另一方面，拓展建筑的功能空间，使其具有更佳的舒适性和更开放的视野。

On one hand, the architect has to retain the main structure and characteristics of the original building, while he transforms the texture and shape of it, and makes it a landmark. On the other hand, he has to enlarge the functional space, and makes them more comfortable with more open views.

GMZY

客房公共阳台区的条形开窗。
The bar-shaped window at the inner side of the public balcony.

整个改造尽可能从周边环境中汲取灵感和用材。设计师认为，旧建筑改造的价值不只在于让空间满足当下的功能需求，更多的是在各种限制中寻求突破，在新旧间找到平衡，赋予旧建筑更为长远的生命周期，让物性延绵，让环境持续。

Most inspiration and materials come from the surrounding environment as planned. The designer believes that the value of renovation is to let the space meet the current functional requirements. Moreover, he seeks breakthroughs within the limits, finds a balance between the old and the new, gives the old building a longer life, and allows the environment to continue.

本页和右页：曦所入口。素白洗墙，山松掩映，门旁仅一＂曦＂字点题。

This page and opposite: The entrance, white wall complements a green pine, and a Chinese character of XI beside the door, highlights the theme.

茶韵墨香——曦所

Flavor of tea with scent of ink – Xi Club

曦所是一个基于兴趣和遵循内心的设计，也是设计者的生活方式和态度的表达。这里整合了茶室、工作空间、雅集……他生活和工作中这些最常态的事和物，体现了他内心的精神诉求。

The designer is the owner of Xi Club himself. Based on his interest, the design totally follows his heart, and expresses his lifestyle and attitude. Xi Club integrates tearooms, a working place, and private collections…which fill his daily life, and satisfy his inner spiritual aspirations.

书案空间。
Writing space.

大厅敞开，靠墙有一张紫光檀明式书桌，线窗稍高于椅背一线，上悬一幅林先生所赠的行草书法《双井茶送子瞻诗》，讲的是黄庭坚送茶给苏轼的故事。这既是老师对学生的祝福，也是对其人生态度的提点。

The lobby is open, and against the wall is an African blackwood desk of Ming-style. A bar-shaped window is a little higher than the chair's back, upon which hangs a piece of cursive calligraphy gifted by his tutor Mr. Lin. It is a poem by Tingjian Huang, which conveys a blessing and a mention from the tutor to the designer.

本页和右页：两幅收藏作品。曦所所陈大部分器物，都是设计师自己多年的兴趣积累，陈列的字画、唐卡等艺术品，多出自名家之手，都是因缘际会，偶得而来。

This page and opposite: Two paintings. Most of the pieces in Xi Club are designer's own collections over the years, and the calligraphy, painting, and Thangka are mostly from famous masters by chance.

右页：定制实木陈列架与藏品。
30 页：茶桌一角。
Opposite: Customized wooden display rack and some collections.
p.30: A corner of the tea desk.

这些艺术珍品与古木、沉香、太湖石、紫砂砚、老坑砚，以及中式家具一起构建了一个精致的人文空间。在这个空间里，可以和朋友饮茶挥毫，也可以把酒言欢，又或者焚香抚琴……
人和物的关系颇为微妙，年月久了，相伴左右的器物，就会成为生活的珍贵记忆，内心的情感寄托。在设计师本人看来，收藏是一种惜物感的延续。他认为生活需要一点精致感和仪式感，在能力之内，选择自己认为最好的，既是对品质的追求，也是对自己的尊重。而展现最好的自己，亦是对别人的一种尊重。
曦所包容所有美好和事物，并期待一切美好的发生。

These art treasures together with ancient wood, agarwood, Taihu stone, as well as purple clay inkstone, old pit inkstone, and Chinese furniture, build an exquisite and humanistic space, where you can have tea with friends, or burn incense and play the zither...
The relationship between people and objects is quite subtle. As time goes by, the objects that accompany us will become our precious memories and affections. In the designer's opinion, collecting things is a kind of spirit of treasuring things, rather than chasing wealth. A refined life requires a sense of ritual. Choose the best within your ability. It is a way of respecting yourself and others as well.
Xi Club embraces all things of beauty, and expects all good things to happen.

人正似松
心美如華

Pay attention to the synchronization of space and
human, and memories of feelings in the space.
Infuse art to life to make the space more artistic. Devote
to creating personalized art space
with vitality and growth.
– Liangfeng Hu

关注空间与人的契合度，
注重空间中的情感记忆。
将艺术融入生活，
让空间更显艺术。
致力于创作具有生命力、
成长性的个性化艺术空间。
——胡梁锋

水墨意境的背景墙。
The background wall in lounge with an ink painting on it.

隐匿市井的百年春秋——大步里院
History hidden in downtown – Dabu Courtyard

大步里院位于宁波市中心，是集办公、美育、设计师买手店、音乐趴房、茶室、咖啡吧、书吧与艺术展览空间的文创园。

设计师通过“新”与“旧”、机能“退化”与自然“包浆”、“闹”与“静”、“市井”与“院景”、“山”与“水”等手法进行设计改造，充分体现了人与自然互融共生的核心哲学。

The project is located in the city center of Ningbo, which consists of office space, beauty care, boutique, music area, tearoom, café, bookstore and art exhibition area.

The architect focuses on the mixture of human and nature, applying the design methods of contrasting "new" and "old", functional "degradation" and natural "patina", "noisy" and "quiet", "outer street" and "inner courtyard", as well as "mountain" and "river".

从玻璃盒子一旁望向书架。
View of the bookshelf seen beside the Glass Box.

老房子的改建是很困难的，设计师需要决定保留哪些、拆除哪些，甚至还要专门邀请专家，协调相关部门，重新开挖管道，把院子的排水和市政管道连上。
当然，改建的过程也有惊喜。在拆除那些已经废旧的隔墙和天花时，房子的本来面貌逐渐浮现，如开放的空间，交错的角梁，还有漂亮的拱券。而随着这幢房子原主人的后人来到这里参观，以及周围邻居老人们的讲述，小院精彩的历史也逐渐清晰起来：这是一幢有着百年历史的民国时期的建筑，1949 年以后原主人把它捐献给了政府，成为行政公署办公的地方；后来，江东区教育局也曾在此办公；这里曾经有个水塘，是周围孩子们夏天嬉水的地方……或许一切都变了，就像当年的小树苗已经变成现在的老槐树一样，但幸运的是，这段历史从掩埋的尘土中“走”了出来，建筑的深度和历史的温度让这个院子有了触动人心的力量。
It is not easy to deal with the renovation of the old house. The architect has to figure out what should be preserved, and what should be removed, and in addition, sometimes to invite experts and related government departments involved to determine how to deal with underground pipes.
Of course, there is a surprise as well. When walls and ceilings are removed, the original appearance of the house slowly appears in sight, such as open space, crossing beams, and beautiful arches. Offspring of the original homeowner and old neighbors also unveils the story of the house. The building can be traced back to the era of Republic of China, lasting over 100 years. Sometime after 1949, the original homeowner donated this house to the government as an office, and later it was occupied by Education Bureau of Jiangdong District. There was also a pond for children to play in summer…It is true that everything has changed, just like the grown pagoda tree, but fortunately, the history of this house is now discovered. The depth of the story and the warmth of the time altogether give it the power to touch people's hearts.

2011 CHINA
STYLISH SPA
HOTEL

左页：玻璃盒子与木屋顶。本页：会议室。
Opposite: View of the Glass Box and wooden roof. This page: View of the meeting room towards the window.

这个蒙尘四年的院子，在一个快要被遗忘的市井角落，以一个有树、有水、有茶、有人的空间迎来她的新生。建筑原有的印迹和人的痕迹，成就了空间的丰满和生机。

The courtyard has been almost forgotten by citizens for four years. But now she is embracing her regeneration with new imges, full of trees, water, tea and people. The original traces also form the richness and vitality of the space.

院子一景。
Part view of the courtyard.

Strictly follow the concept of "making others successful before achieving own success". Provide clients of different fields with customized design services covering planning, architecture, landscape, indoor design, exhibition, etc. The design emphasizes details and quality and spatial experiences. Modern approaches are often used, while eastern spirits are highly recommended. Works are modern, natural and humanistic.

– Caifu Li

秉持"立己达人、成人之美"的哲学态度，
以规划、建筑、景观、
室内与陈设等全面的设计系统，
为不同领域的高端客户提供特定的设计定制服务。
设计创作讲究细节品质、注重空间体验，
善用现代手法、崇尚东方精神。
作品呈现当代性、自然性与人文性。

——李财赋

本页：绿意环绕的露台区。右页：3D 模型图。

This page: The terrace surrounded by green. Opposite: 3D model of the Hui House.

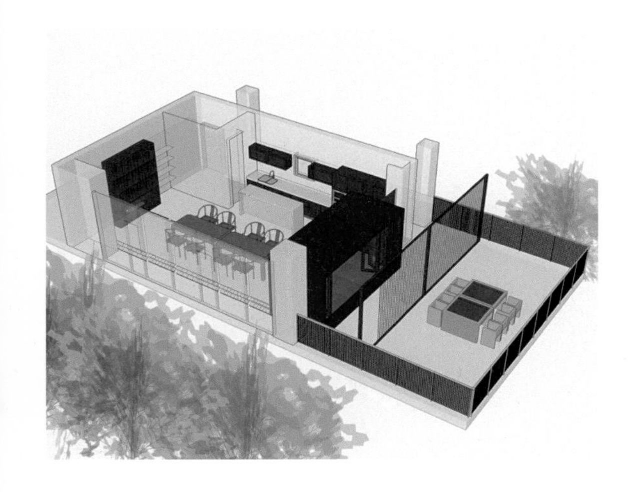

朝九晚五里的诗和远方——慧舍

Poetry and dream in working time – Hui House

改造后的慧舍，集茶室、艺展、设计于一体，是一个能够体现江南生活美学的空间。

这里不仅是招待客户的地方，还是一个以茶会友的雅舍，也可以是以烧烤为主题的聚会现场，更是可以共享的“诗和远方”。

整个功能区主要分为室内空间和露台空间，中间用一个抬高 300 毫米的榻榻米茶室过渡。

This project, which consists of tearoom and meeting room, demonstrates the beauty of the regional lifestyle of Jiangnan. It is not only a greeting area for clients, but also a tearoom for guests and friends. What is more, it can turn to be the scene of a party with barbeque theme, as well as a place to share “poetic life”.

The whole place is mainly divided into two parts – the indoor space and terrace space. They are connected by a tatami tearoom that is lifted by 300 mm.

右页：慧舍入口。44~45 页 : 铁盒子内的茶空间，也是室内外的过渡空间。

Opposite: Entrance of the Hui House.
pp.44-45: Tea room within the iron box.

本来面对露台的墙是封闭的，在设计师的努力下，大楼的主人最终被打动，同意把墙打掉，把满眼的绿色引入空间。设计师把一个方形的铁盒子置入茶舍和露台之间，取代了原先的封闭墙面，通过可以完全打开的折叠门，完成这个空间的延续。

慧舍的周边被大树环绕，坐在这里喝茶，人自然便得到了放松。若谈笑间，风满座，得水声入耳，岂不妙哉？为此设计师养竹造景 , 用近 3 米的旧马槽、旧时的石磨和当地的竹筒，做成水循环，空间里有了潺潺水声，便又添几分生趣。再配上有几百年历史的石凳，让全新的空间里有了历史的痕迹。

The wall opposite to the terrace was originally closed, but the building owner eventually agreed to remove the wall and introduce the green into the building after being persuaded hard by the architect. A metal box was put between the tearoom and the terrace to replace the closed wall. After folding doors were added to the metal box, this space is now continuous.

The building is surrounded by plenty of trees. Whoever sits here is relaxed when drinking tea, with gentle breeze fanning our faces and clean sound of water ringing. The architect uses bamboo to create the landscape with a 3-meters-long old manger, an old-fashioned stone mill and local bamboo tubes. A water circulation is thus established. When water is involved in the design, more fun is created. All these things fit well with the stone bench that has a history of several hundred years, imprinting the trace of time.

从室内望向露台。
Looking toward the terrace from the interior.

整个改造是一个整理思路、重新出发的过程。在繁华的都市里，辟得一域，诠释着都市可同大自然亲密相容，诗和远方亦可以在朝九晚五之中。

壁炉区，与有中式意境的茶室区形成中西合璧的氛围。
The fireplace, contrasting with the Chinese-style tearoom.

The process of renovating involves a concept of sorting and restarting. Establishing an area in the metropolis to show that the city can mix with the nature well, and that poetic life can also mix with the routine well.

室内全景。

The overall view inside the Hui House.

What I devoted to doing,
is infusing the space with emotions and creating unique works. The space with wisdom cannot exist without life, culture, or art. I know the regional culture of Jiangnan very well, so I usually use feelings of literati from there as the base to create eastern atmosphere.
The design should not only root in tradition,
but also forecast the future.
It should be unique and creative.
The core of design is craftsmanship.
– Gaofeng Pan

我一直坚持在做的，
就是赋予空间感情，
做独立有创性的作品。
有智慧的空间离不开生活，
离不开文化，离不开艺术。
深植江南文化，演绎东方意境，
以江南文人情怀为轴，
表明设计既要根植于传统更要展望未来，
坚持设计的唯一性，独创性。
设计的核心价值就是拥有一颗为设计的匠心。
——潘高峰

休闲区。

Leisure area.

庭院深深深几许——楠山南

Courtyard in loneliness and distance – South of Mount Nan

这是一个老房改造项目，保留了老房子原有的建筑格局，融入了现代家居材质，吸纳了东方水墨留白意境，游走于传统与现代之间，不张扬，不矫饰，期冀彰显一种质朴的审美境界，营造出一个有归属感的、古朴大气的静美民宿。

This is a project of renovation. The structure of the old construction is preserved, while modern materials are used. Eastern aura is infused into the project, and the architect wishes to demonstrate a simple and pure beauty, and build an elegant rural inn with the sense of belonging.

本页：茶室。右页：公共活动区。
This page: The tea room. Opposite: The lobby.

保留了中式风格的传统雕刻案几、木雕门窗、古玩装饰；白墙、原木、国画、小绿植，使得整体空间呈现原生态感。

Traditional Chinese style architectural parts are preserved in the project, such as carving tables, woodcarving window frames, and Chinese style antiques. The whole space shows the sense of nature with white walls, logs, Chinese paintings and some green plants.

本页：公共空间壁炉区。右页：客房。

This page: The fireplace area. Opposite: One of the guestrooms.

房间保留了原有的端正稳重的房梁，提升了空间的层次感。采用传统榫卯工艺的原木床榻和电视柜与素色窗和单人沙发，是一种中西生活的碰撞。

The original straight and solid beams remains at the same place. The bed and TV stand connected by mortise and tenon joint which is a traditional woodcraft, contradicts with the modern single-colored curtains and one-seat sofas, showing contrast of lifestyles of western and eastern.

走廊。
The corridor.

Learn to be calm and learn to listen.
Only when you listen carefully without any desire,
can you notice the poetic part in words.
Learn to listen to the hidden contents
between people and space.
Thus, one can truly understand
the demand of user of the space.
Architects should use their calmest hearts to
experience, then they can conduct the true design.
– Zhiming Liu

做设计这些年，
让我历练了从浮躁的心到有耐心到平静心，
对事物的看法也似乎简单清晰了。
让心静下来，安静地聆听身边的人和事。
只有不投射任何欲求，
才能听到话中的诗意，
听到人与空间的微妙，
也更能深入地了解人对空间的需求。
用最平常的心，去成就设计对人与生活的意义。
——刘志明

办公空间
Office Space

前台一隅。

A corner of the reception hall.

简的艺术——杭州振邦律师事务所办公室

Art of simpleness – Office of Zhenbang Law Firm in Hangzhou

设计师相信美好的空间并不需要过多的装饰，大巧不工。设计的内核是解决空间的问题、功能的问题，以及空间与人的关系。项目总面积约 1100 平方米，占据办公楼一整层的空间。平面呈“回”字形布局，由电梯厅南部进入接待大厅，设计按照律所的办公和业务流程逆时针布局不同功能空间，实现员工之间的相互协调配合。

The architect believes that decent space does not need to be overdecorated since less does more. The true designing is to build better space with better function and to establish better relationship between space and human.

With total area of approximately 1,100 m^2, the project occupies a whole floor. The plan looks like a square block with a square hollow, with the reception hall in the south of the elevator. The functional area is planned in counter clockwise according to the demands of attorney-related business, contributing to the cooperation among employees.

本页：转角空间。右页，上及下：光影之约。
This page: Corner space. Opposite, top and bottom: The dance of light and shadow.

在本案中，设计师首先以简单的几何线性梳理空间，构建空间的秩序和形式的有序化；其次从甲方的行业属性出发，以其工作流程为规划设计核心，结合企业的员工构成、客户构成和业务构成，充分发掘实际需求，并在原建筑结构分析的基础上，规划空间的功能布局和行为动线，通过对项目内外环境关系的处理以及对人、光照、通风和使用效果的考虑，实现空间的构成。

In the designing process, the architect first uses simple geometric shapes to build space order. Then, the industry's attribute is taken into full consideration: working requirements are put on the first place, and employees and clients are considered as significant aspects as well. The architect discovers what the space truly needs, and then plans functional layout and circulation design based on analysis of the original building. Eventually, the project is established through the communication of inner and outer environment, and through the usage of natural lights and air ventilation.

左页：玻璃隔断细部。本页：会议室的一抹绿。

Opposite: Details of the glass partition. This page: A green plant in the meeting room.

通过超大的落地玻璃窗，可看见钱塘江的优美景观。简约的设计美学使得这里呈现一种艺术空间般的简净气质。设计师认为，高级的审美，是对自然的尊重和热爱，是对体感和情感的体会和感触，是在简约中发现事物变化的美。

Water view from QianTang River is introduced into the building through large floor-to-ceiling windows. A simple design gives out an artistic atmosphere to the project. The architect believes that true aesthetic is to respect and love the nature, to perceive and understand feelings, and to unveil the beauty of changing through simpleness.

前台区，高高的挑空，宽阔而亮堂。

View of the reception hall, with great height, spacious, and bright.

主接待大厅充分发挥原空间的结构优势，挑空的空间具备综合的功能。这里不仅是展示企业形象和接待客户的空间，更是企业团建、灵感与创意相互交织碰撞的复合功能空间。在给予客户礼遇的同时，赋予员工更多的温度。

The reception hall is not only the brand image of the firm and greeting space, but also a palce with multi-function of communication and creation. As a result, this hall is designed to fulfill various usage with atrium, a major update for both employees and clients.

报告厅的阶梯式盒子空间。
The stepped seating space.

自然与秩序——玉米之家
Nature and order – Office of Corn Design

随着环境的影响力逐渐渗透到人的行为中，传统的办公空间已经无法满足当代多元化的工作需要。以年轻团队为主体的玉米设计将人与环境的界限逐层打破，重新解读空间的内在表现形式，在保留办公功能的基础上，用自然纯粹的语言糅合空间的个性元素，聚焦每一位员工的情感体验，亲手打造了这处自然与秩序并存的“玉米之家”。

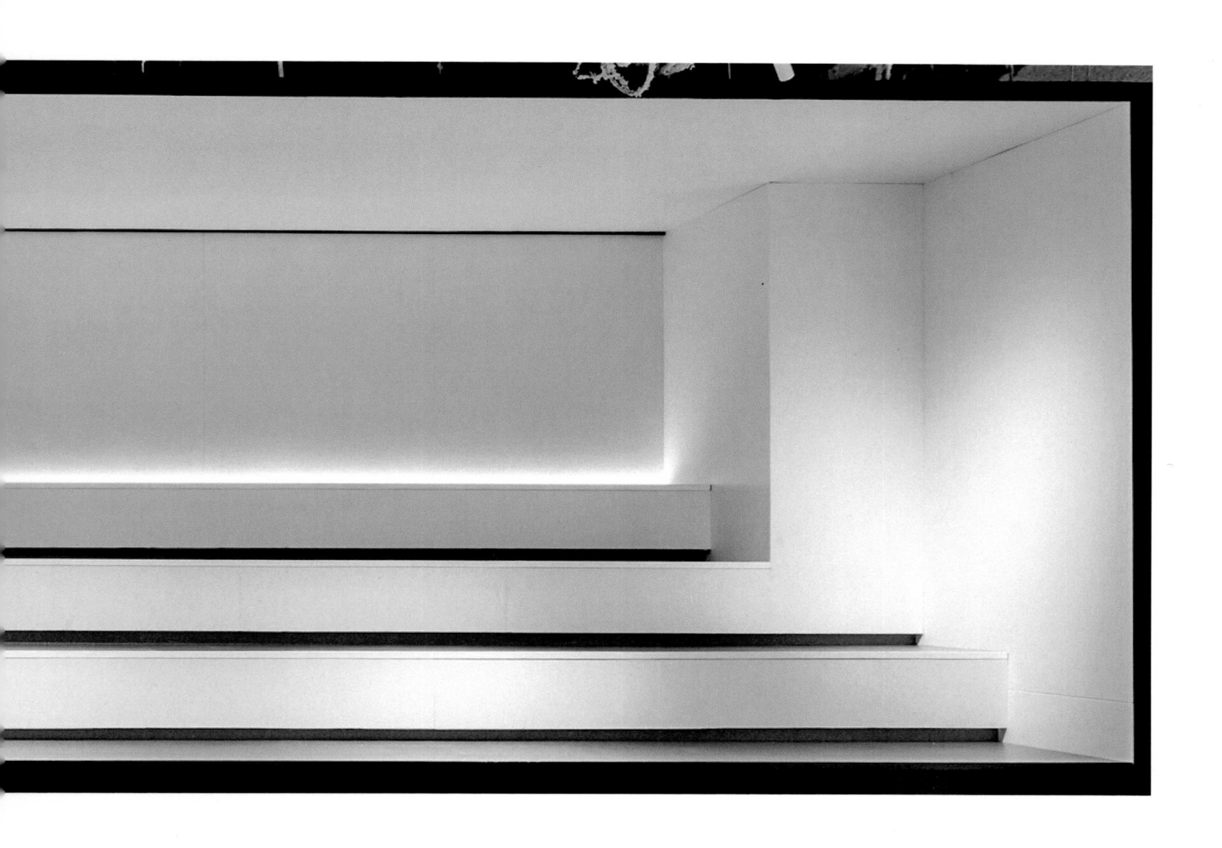

Traditional offices can no longer satisfy modern and diversified working needs as the workplace environment is influencing people more and more. Corn Design, full of young teams, breaks boundaries between human and environment, redefines the true form of space, and finally builds the corn family that has both nature and order. With practical office functions remained, this project is built with natural and pure designing language, mixed with unique elements, and infused with emotions from all employees.

开放的报告厅空间。

The wide opened reporting hall.

公共活动区作为人际交流、思想碰撞之地，设计师隐去了墙体的阻隔，使接待区、报告厅、会议室融为一体，在递进与交互中演绎多层次之美。报告厅以独立的内建筑为框架，自成一体，静伫其间。阶梯式座位，随意摆放的各色柱形凳，营造了轻松惬意的氛围。

As the place to exchange thoughts, the public zone needs a considerably large area. Consequently, the designer breaks the wall and brings the waiting area, sharing hall, and conference room together to consist a whole space, conducting the beauty of multi-layers in progress and interaction. There is an independent box in the hall, embracing the space with open stances. Seats are set like staircases, and stools of various colors are set in front, forming an easy and neat atmosphere.

左页：植物墙近景。本页：从会议室里面望向报告厅，可见前方植物墙和接待区。

Opposite: Close view of the plants wall. This page: The plants wall and reception area can be seen in reporting hall seen from the meeting room.

绿植墙和接待区。
The plants wall and the reception area.

接待区鲜活的绿植强化了空间的自然肌理，深浅不一的植被覆盖了整个墙面，自下而上仰望，绿植墙宛若绿色瀑布倾泻而下，带来强烈的视觉冲击感。屋顶的镜面装置延伸了绿植墙的广度和深度，现代软装与原始自然的交织，让人处于大自然的包裹中，感受随性舒适的现代办公体验。

There is a green plants wall in the reception area, deepening the natural texture of the space. The whole wall is covered with green color of different shades, forming a "waterfall" to enhance the visual impact. The mirrors on the ceiling enlarge the green wall. Thus, modern decoration and primitive nature meets together, creating a natural atmosphere and comfortable office experience.

本页：阶梯式座位近景；右页：工作区的隔断。

This page: Close view of the stepped seating; Opposite: Partition in working area.

员工区以黑白灰为基础色调，用简约的线条营造出纯粹契合的空间感受。减噪的成品方块地毯结合环保有机棉天顶修饰了整个办公区，原始粗粝的元素将办公区低调的风格再次凸显，在快节奏的都市办公环境中注入润物细无声的自由与舒适。员工可以各凭喜好装饰自己的小天地，积极参与到空间的创造中，从而传递愉悦和灵感，迸发出无限的创意。

The employee zone uses black, white and grey as base colors, creating pure and compatible feelings with simple lines. Square carpets with noise reduction function and organic cotton ceilings decorate the whole office area. Primal elements highlight the low-profile style of the office space, and the fulfilled room with liberty and comfort quietly. Employees can decorate their own office space freely, join in the creating process of the space, pass joy and inspiration to each other.

本页和右页：入口和靠窗的墙面。
This page and opposite: The entrance and a wall near the window.

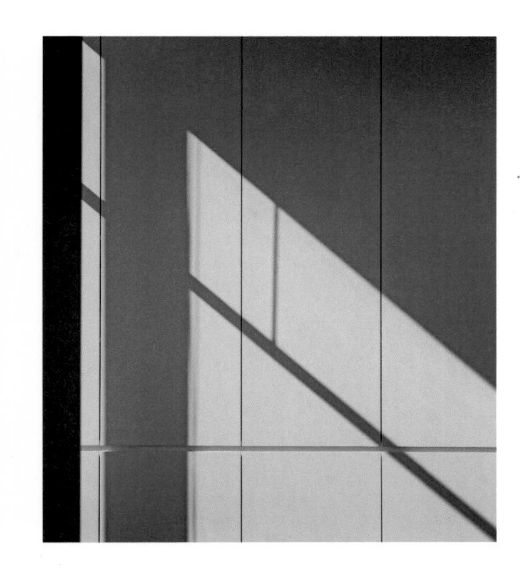

城市之光——浙江欧硕律师事务所办公室

Light of city – Office of Zhejiang Oushuo Law Firm

设计注重透明度、轻盈感和自然采光，开放的空间与极简的材质结合，完成对功能、感官的设计需求。同时借由无遮挡的空间界限，将城市天际线和内部空间的交互限制解除，让城市天空与四季变化成为空间的景观和陈设。

Transparency and lightness of the space as well as the use of natural light are highly emphasized by combining the open space with extremely simple materials to meet design requirement of function and sense. Meanwhile, unshaded spatial boundaries release the restriction of city skyline and inner space, thus taking the sky and changing seasons as a part of the view.

本页和右页：办公室的窗景。
This page and opposite: Views through windows in the office.

光是这个空间的记忆点，让光与空间直接对话，通过晨曦、白日和夕阳，结合冷静的色彩，简洁的线条，让直白的现代空间自然拥有丰富的表情，从而营造了四时不同的办公体验。

Light is the key to the space. Letting light in, whether it's at dawn or dusk. When light meets the cool and clean color and lines, the modern space thus has a rich experience of different seasons and feelings.

独立洽谈室。

The independent negotiation room.

阳光走廊。
The sunlight corridor.

自然光透过玻璃幕墙进入空间，满屋光亮，那些经过玻璃折射的光，投射在墙上的线条，铺陈在地毯上的柔和光晕以及反射在石材上的倒影，相互交融渗透，在一天的不同时段，自然地变化着，带给使用者不同的体会和感受。
模块化的办公空间，多功能的办公空间组合，服务于当下和多变的未来。

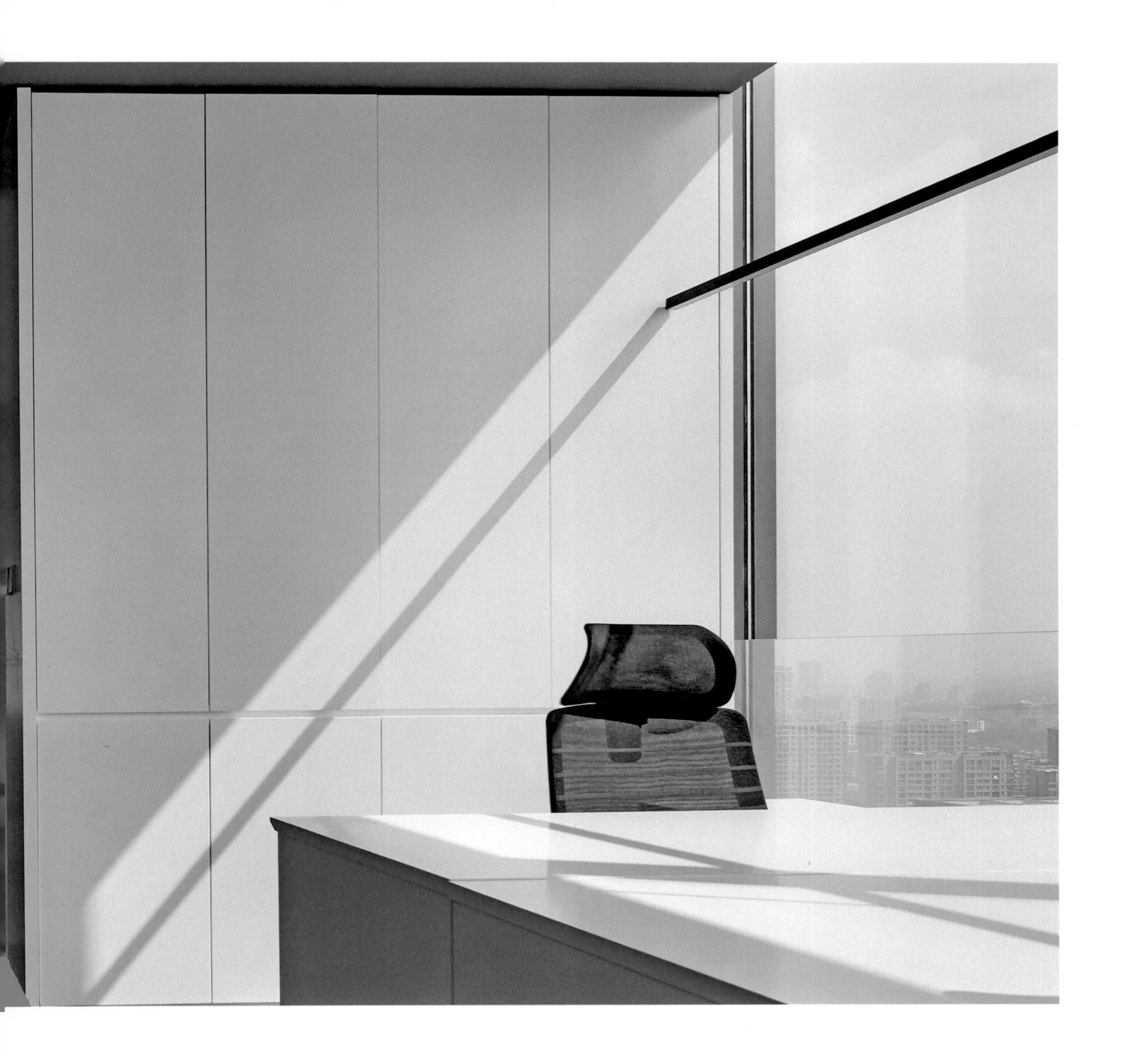

Natural light brightens the inner space through the glass curtain wall, being reflected and mingled by the wall, blanket and stone. In different periods of time, the natural change brings different sense and feelings to the people in it.

Different kinds of office design serves the present and changing future.

开放的办公区。
The open working area.

冷静、睿智、公平与开放，设计师以黑白灰为主要色调贯穿整个空间，将律师的职业精神融入办公环境之中，并致敬这群用理性和法律守护着城市秩序的人们。

Black, white and gray run through the whole space as main color by which the designer wants to blend lawyer's professional spirits into the office, that is calmness, wisdom, justice and openness. The design is also in honor to the men and women who guard the city with order and law.

这个城市，总有一些人，
用自己的智慧和态度，
守护着城市之光。

There are always some people in
this city guarding the city's light
with their wisdom and attitude.

左页：大堂区。
Opposite: The lobby.

思·源——宁波旷世智源办公室
Thinking & origin – KWUNG'S Ningbo Office

设计是基于本质的理念升华，经由思考的过程，让天马行空的想象，从图纸到落地状态。它源于面对未知、新生时的憧憬，需要多元的灵感与支撑。

宁波旷世智源，作为文化创意企业，专注于工艺家居饰品的研发、设计等系列工作。在发展理念中，时刻强调“品牌、创意引领市场”。为了突出艺术特性，在室内部分，设计师着重打造别具一格的办公风格，努力营造不受限、多可能的办公空间。

To design is to sublime. After the process of thinking, the conceptual and initial imagination are transformed from drawings to constructions. It originated from anticipation of the unknown and newborn. Diversified inspiration and supports are needed during the design process.

As a company in cultural and creative industry, KWUNG'S focuses on researching and designing artistic decorations, and always emphasizes on "leading the market with brand and creativity". To highlight art property, designers set a unique office style in indoor space, trying hard to establish an unlimited and multi-usage office space.

将黑白对比的不规则元素加入到中性灰的高级感中，让开阔的空间既有工业风的冷酷沉稳，又不失现代简约的个性。

Adding the irregular elements of black and white to the advanced sense of neutral grey, granting the wide opened space a cool style, while keeping it modern and simple.

本页：会议室；右页：办公区。
This page: The meeting room; Opposite: The working area.

原木和绿植的点缀，以自然的色彩与肌理，为空间注入生机。楼梯曲折却不弯卷，用干净利落的线条，使常规楼梯多了一些新意。

Wood and plants are used in the office to enrich vitality of the room with natural color and texture. The stairs seem folded but not curved, adding more refreshing elements through clean and neat lines.

在办公区的过道处，抬眼可以见顶。设计师不对高挑的层高多加阻碍，而是用大间隔的短梁，对楼层进行区分，使得空间通透、敞亮，避免产生压抑感。

Rooftop can be seen in the hallway. The architect doesn't set any obstacles in the space with great height. Instead, several short beams are used with some distance among each other to separate every story. The space is thus more clear, bright and without visual pressure.

右页：大堂区的柱体。

Opposite: Pillars in the lobby.

整齐的桌椅排布和近乎统一的黑白灰主调，能够呈现办公区制度化的规矩、条理。而通过多边形门框以及三角色块等几何元素的介入，又能使其氛围更具灵动性。一些大面积或细节处的鲜亮色块，有助于员工在更为轻松舒适的环境下，寻找思维的突破口。

纵观整体，本案设计中少有门与隔板的出现，或是以玻璃为墙，尽可能在保留必要私密空间的同时，增强工作中的流动性，使每个人都能在局部领域中，实现自我的存在价值，找到归属感与责任感。并且，更有效地促进员工间的相互交流，以提高团队协作及凝聚力。

The main colors used in the office are black, white, and gray. These colors, together with tidy desks and chairs, express a well-organized office area. On the other hand, polygon-shaped door frame, triangle color blocks and other geometry elements are introduced to form the lively atmosphere. Some color blocks with large area or bright color in details are to help employees with their thinking process in more comfortable environment.

All in all, there are few doors or wall plates in the room. Instead, glass is the main material for walls. This is to enhance mobility in the work while still maintain a certain level of privacy. Every employee can realize self-worth and find sense of belonging in a certain field. What is more, communication between workers is encouraged through the design method, as well as the teamwork capability.

空间和情绪——可瑞舒适家杭州总部

Space and emotion – Co-Real HOME SMART in Hangzhou

可瑞舒适家杭州新总部大楼，位于杭州西湖区古墩路上，由一个上下两层的简易厂房改造而成。设计师一改原建筑的脏乱形象，打造出一个以挑高中庭为核心的体块分明的白色盒子空间，使其独立于周边环境。

傍晚时分的建筑外墙。
View of the facade at dusk.

Located in the West Lake District of Hangzhou, the new headquarter of Co-real Home Smart is a renovation of a simple factory. The designer changed the dirty image of the original building to create a white facade with distinct volumes around a central courtyard, making it independent from the surroundings.

本页：室内的自然光表情。右页，上及下：大厅局部的红与白。
This page: Interior expression of natural light; Opposite, top and below: Red and white in the main hall.

中庭的顶部，以一个有秩序感的造型，消除了大部分的太阳热量，使室内整体空间常年保持恒温恒湿，减少外部影响。同时，经过计算和设计，巧妙地让一部分光线进来，光线随着太阳位置的变化而在空间中移动，好像是可以读取的天然时针，让空间变得灵动无比。

The top of the atrium is designed to follow the beauty of order, while the atrium can also eliminate most sun heat, bringing the indoor space homeothermic and homeo-humidity atmosphere for all year. Meanwhile, according to careful calculating and designing, the sunlight is introduced with exact amount and exact angle, creating a natural time pointer when the sun moves, adding a sense of ingenious variety to the space.

大厅全景。
Overall view of the main hall.

整体空间简净是第一语言，探究一种“白”的实体，感受白的纯粹，并在白中大胆地融入红色，希望调动空间使用者积极乐观、热情主动的情绪。

The top design language used in this space is simpleness. By studying the reality of "white", feeling the purity of white, and infusing red to white innovatively, the designer wishes to induce optimism and enthusiasm of users.

中庭盒子。
The atrium.

大楼一层设有产品展厅和可瑞舒适家各系统的智能体验样板间，二层为综合办公区。

The first floor of the new building is equipped with a product showroom and the intelligent experience sample room of each system of Co-real Home Smart. The second floor is a comprehensive office area.

仰视中庭盒子。

Upward view of the atrium.

红白黑办公桌椅。

Working desk and chairs in color of red, white and black.

I am interested in interaction designing, and prefer plugging in the design concept into business value. Focusing on discovering inspiration, connecting thesis concept, construction space, material, and light and shade together to establish a clear theme and aesthetics of the space.

– Yun Mao

我喜欢空间上的交互设计，
并将设计理念嵌入到商业价值中。
乐于不断发掘创作灵感，
将主题理念、建筑空间、
材料材质、光影关系融合在一起，
赋予空间明确的主题属性和美学气质。

——毛赟

拒绝平庸，在尊重使用者需求的基础上探寻着空间更大的可能性。

Refuse to be ordinary. Every part of design is always to seek for a larger possibility of space usage based on true demands from users.

左页：过道，尽端是经典的蓝白红几何图案。112~113 页：等候区。

Opposite: The corridor with classic bule, red and white pattern in the end.
pp.112-113: Waiting area.

新锐活力——博洋前洋 26 联合办公

Dynamic youth – Beyond Qianyang 26 Joint Office

博洋集团前洋 26 联合办公位于宁波江北电商经济创新产业园区，是集理想工作与综合服务于一体的年轻化生态。设计师以“嘻随之趣，勤思之态”为设计理念，希望打造一个垂直办公的活力社区，用模块的概念将传统办公模式与未来办公相结合，从而产生一个“新锐、活力、年轻”的富有生命力的办公空间。

Qianyang 26 Joint Office of Beyond Group is located in an e-commerce economic innovation industrial park in Ningbo, with a young ecology integrating ideal work and comprehensive services. Based on the design concept of “following the interest and thinking frequently”, the designer creates a vertical office dynamic community, combining the traditional office mode with the future office with the concept of module, thus creating a “new, dynamic and young” office space with vitality.

左页：转角楼梯。本页：公共开放空间。

Opposite: Staircase in the corner. This page: The public open area.

安静独立的空间带来人性化的舒适体验，模块化的家具组合活跃了整个空间构成。转角处，一抹深蓝穿越在纯净的白色空间里，打通了上下楼的连接，建立了垂直生态，像是有生命一般缓缓向上延伸。

The quiet and independent space brings comfort, and the modular furniture combination activates the entire space composition. At the corner, a touch of dark blue travels through the pure white space, connecting the upper and lower floors, establishing a vertical ecology, which slowly extends upwards like something living.

等候区的金色天顶。
The golden ceiling in the waiting area.

等待区金黄色的顶部星光闪耀，独立成一处存在感极强的天地，斑斓的软装像年轻人活跃的思维，随时随地迸发灵感，向未来铺设无限可能。

The golden stars on the ceiling of the waiting area are shining into a world with a strong sense of existence. The colorful soft outfits are like the active thinking of young people, bursting with inspiration anytime and anywhere, and laying infinite possibilities for the future.

私密的洽谈区。
Private meeting room.

设计师针对性地赋予不同场景最舒适的办公环境。人与空间、人与人的关系变得紧密而有趣。流畅动感的布局激发人们去探索、发现、遇见。

The designer gives the most comfortable environment in different scenarios. Thus, the relationship between people and space becomes closer and more interesting. The fluent and dynamic arrangement inspires people to explore, discover and meet.

" 盒子 " 空间。
“Box” space.

联合办公空间结合时代美学和形式美感，满足了专注与休闲、站立与放松的多重办公需求，使讨论与专注在日常工作中完美切换，从感性和理性的视角深入细腻地表达了空间情感。

The co-working space combines the aesthetics of the times and the aesthetic feeling of the form, and meets the multiple office needs of focus and leisure, standing and relaxation, making discussion and focus switch perfectly in daily work, and deeply and delicately expresses the space emotion from the perspective of sensibility and rationality.

办公室混凝土柱子上绘画的艺术造型。

Painting on the concrete column in office.

隐于效率的诗意——集艺办公室

Poetry hidden in efficiency – Jiyi Office

这并不是一个为了好看而设计的办公空间，也不是纯粹简化的高效办公区，它更多地是从公司发展和员工需求出发，兼具功能和精神需求的空间。

This is not simply about good-looking, nor is it purely about efficiency. The space has to meet both functional and mental requirement to serve the employees and company development better .

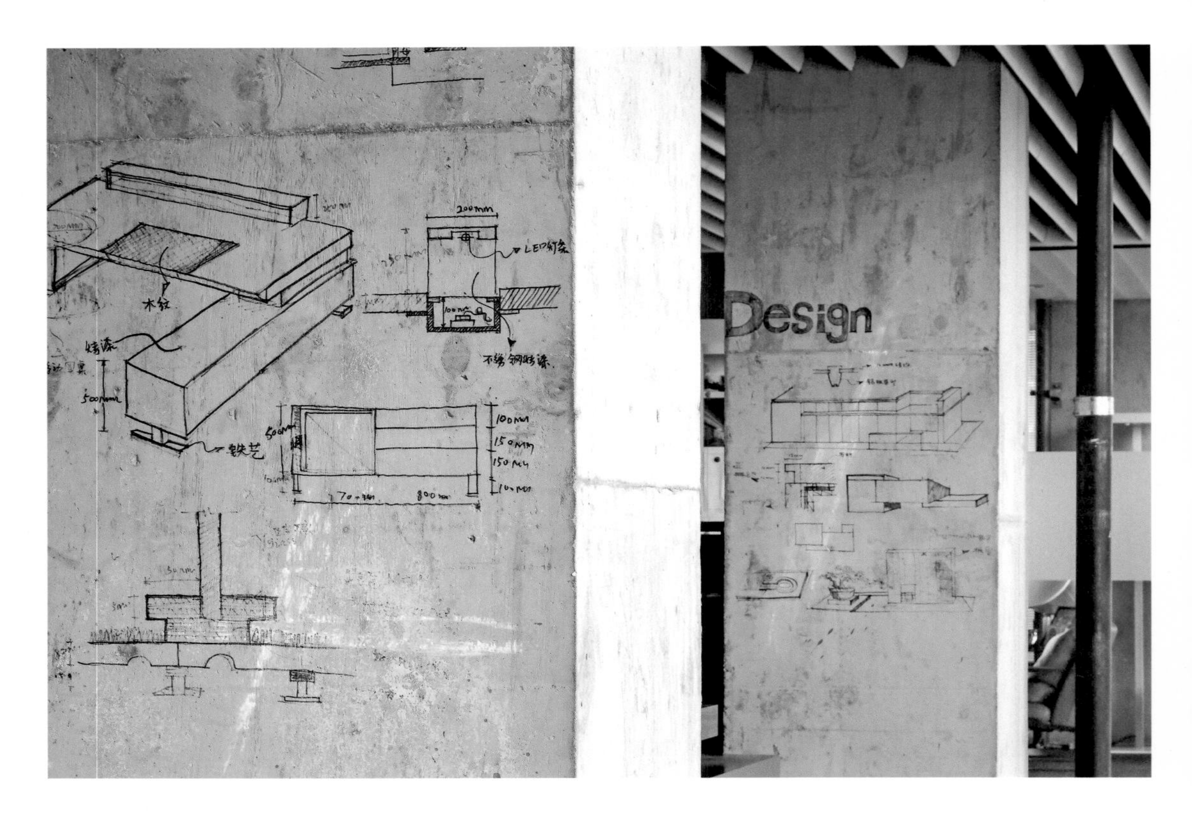

办公室混凝土柱成为方案构思和演绎的图板、灵感的集聚地。
The concrete columns become the drawing board for the scheme conception and interpretation.

设计是从办公座位开始的，从模块设计到动线布局，量身定制了一个内外统一的定制型办公空间。空间整体呈现一个冷灰色基调的轻工业风设计，设计中保留部分混凝土结构，减少墙体阻隔。在隔断上用玻璃材质，使得自然光源得到了较充分的利用。细节处用粗犷材质和细腻材质衔接，与人接触的，必然温和细腻；仅止观赏的，不妨博大粗犷。

The designing process starts from the office desks. From model design to circulation layout, architects have customized a unique office with consistence of both inside and outside. With some concrete constructions preserved, the place is mainly designed in industrial style with grey and other cold colors. Certain walls are removed, and glasses are used to separate each space to make big use of natural light. Therefore, the space brings better vision and feelings, and becomes more environmentally friendly. Rough materials are used together with fine materials, while the former is for viewing and the latter may be touched by users.

上：办公室场景。下：门把手和办公桌的手稿，这里的办公家具都由设计师设计成稿后定制而成。
Top: Whole view of the working area. Bottom: Sketch of the handle and desk, the furniture here are all designed by the architect.

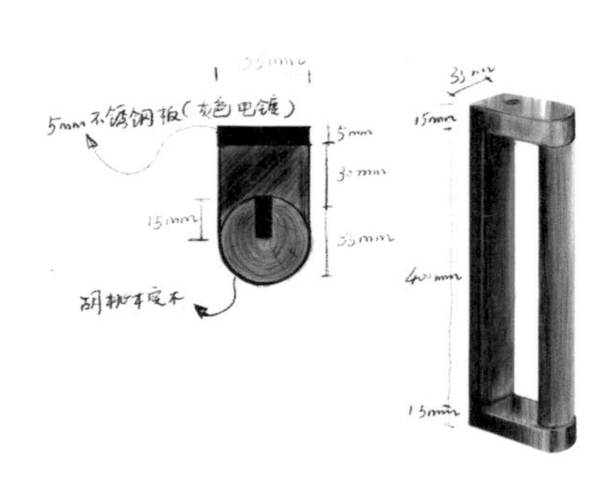

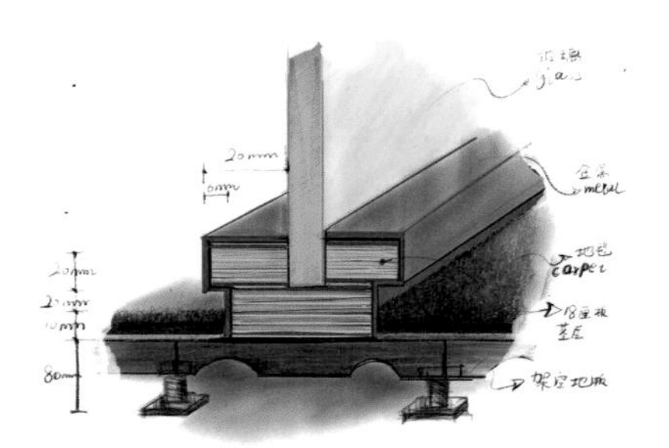

本页，上及下：办公桌的实景和草图。右页：会议室。

This page, top and bottom: The office desk in use and in sketch. Opposite: The meeting room.

會議室
JIYIDESIGN

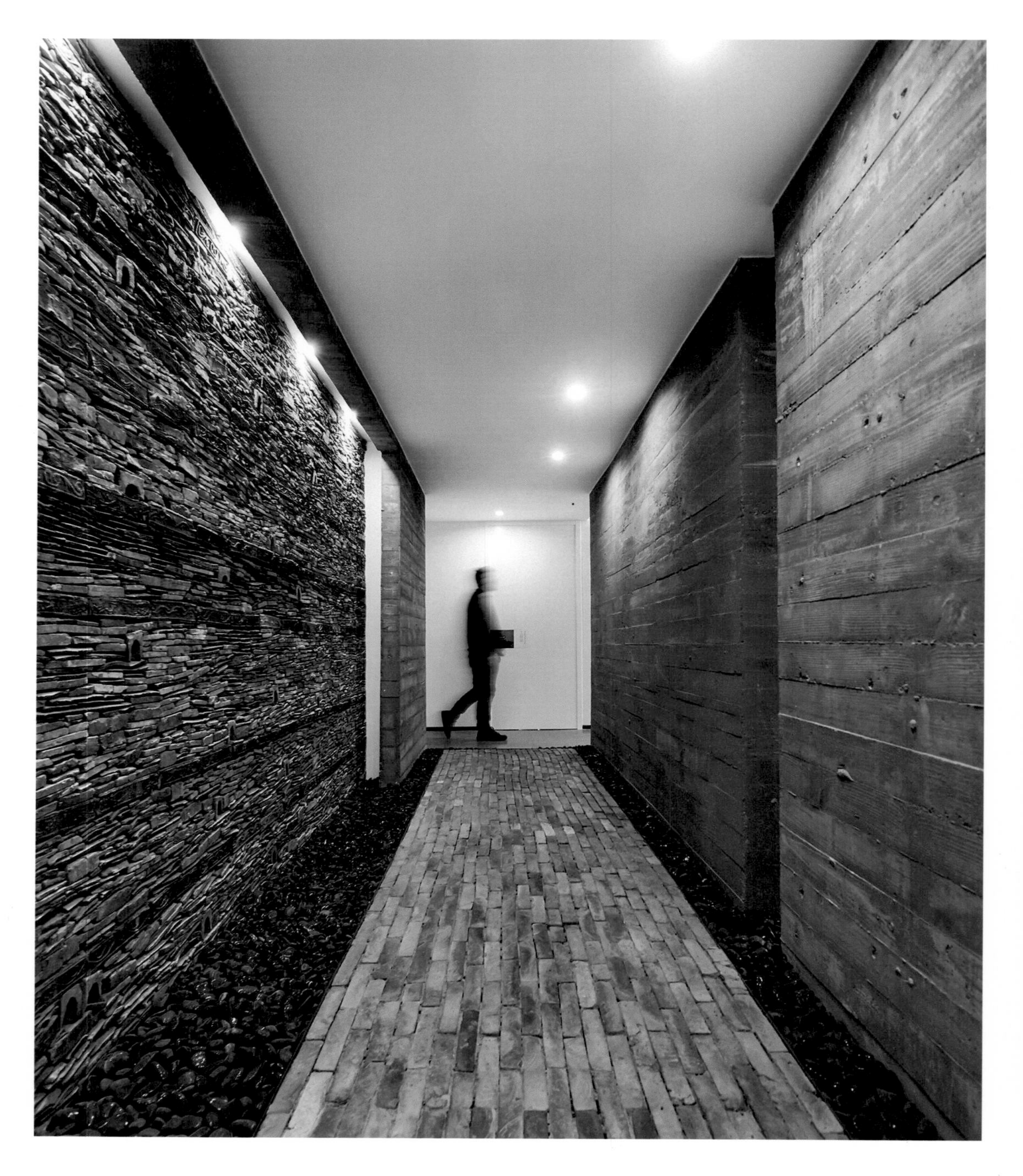

本页：入口走廊。右页：走廊瓦坯墙近景。

This page: The corridor in entrance. Opposite: Close view of the brick wall.

生长的空间——浩然办公室
Growing space – Haoran Office

返璞归真的原材料会随着时间流转而产生不可预测的变化。这些有生命力的材料是运动着的，例如现浇混凝土、老青砖、水磨石地面，它们给这个现代方盒子的办公室注入了温暖人心的温度，并赋予这个全新的现代化办公空间一种安定沉稳的力量。

Raw materials can change over time unpredictably. They are growing and infusing the office, a modern box with a heartwarming atmosphere is warming the office. These livable materials such as cast-in-place concrete, old bricks and terrazzo floor, give the new modern office space a steady force.

由钢索拉伸的楼梯。
Stairs fastened with metal wires.

空间平面规划以区域布局进行分隔，利用材料特性的变化拓展了空间的纵深面积。办公空间的长吊灯让空间更有延展性，同时将导视系统融合在空间里。楼梯采用钢索拉伸的方式，小构件对空间产生了大影响，仿佛有一双结实有力的工匠之手，轻缓而又细致地为你所在的空间护航。设计旨在营造一种安静而不容忽视的力量。

The space is divided by area allocation. Material properties are used to enlarge the space. The long hanging lamp in each office makes the room more flexible, and meanwhile it mixes the signage system into the space. All the stairs are fastened with metal wires. Small components impact the space strongly, creating a feeling that the solid and sophisticated craftsmanship reappears in this space and protects the user with care and gentleness. The designer wishes to establish a quiet yet eye-catching power.

商业展示
Shop & Exhibition Space

左页：从门洞望向就餐区。本页：门店标志，由本案设计师创作。

Opposite: View of the dinning area from the doorway. This page: Logo designed by the architect.

远古印记——烤古烧烤

Ancient imprint – Kaogu BBQ

烧烤所具有的文化属性，一方面是它包含的物质的原味及粗鄙化即食效果中对惯常饮食的反叛，另一方面则是它受到人类远祖在渔猎时代的饮食记忆符码的认同。

Barbeque is a culture, that contains the original taste of the food and rebels to the common meal in the vulgarity eating method. On the other hand, it is also a symbol of fishing and hunting culture in ancient times.

本页和右页：从入口前往就餐区的通道，仿佛通向远古的时光隧道。
This page and opposite: The passage towards dinning area, as if a time tunnel to the ancient.

猛犸象曾是石器时代人类重要的狩猎对象，在许多洞穴遗址中都曾出现。石器时代人类的狩猎对象和生活场所，形成了设计的创作主题。

The mammoth was once a critical hunting target in stone ages, and things related with it appear in many caves and ruins. Therefore, our theme is about the living and hunting places in stone ages.

旧石板、岩壁、工地灯，整个空间用材简单。

Simple materials are used in this place, such as old stone boards, rocks, and lights of construction sites.

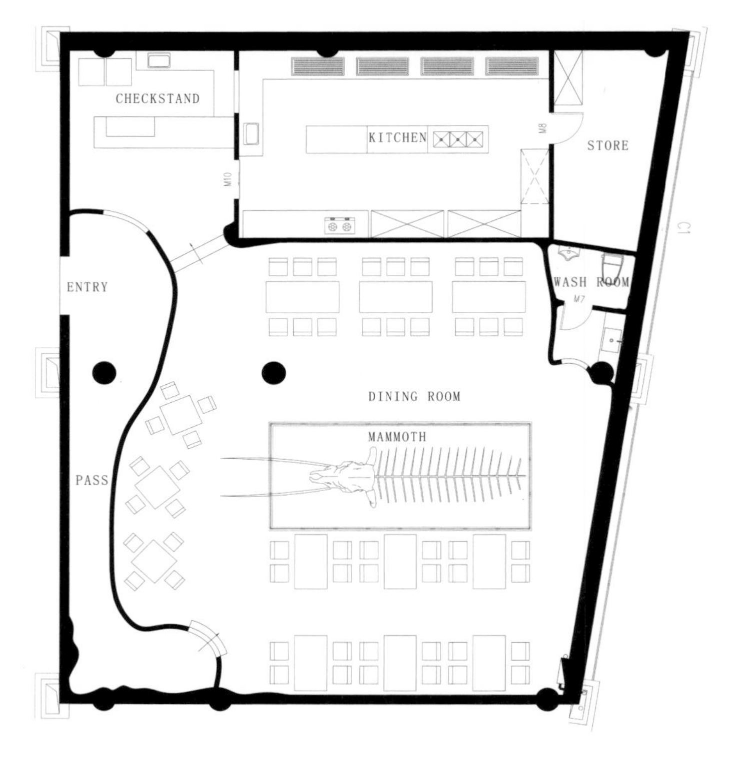

平面图（比例：1/200）
Plan (scale: 1/200)

洞穴连着洞穴，大洞连着小洞，

空间与空间之间产生许多凹陷。

暧昧的秩序、分离又连接的空间、基于精密计算的随意性，塑造出我们想要表达的效果。

猛犸象化石雕塑，既渲染了氛围，又表达了主题。

Sculpture of the mammoth fossil . It not only sets the mood, but also expresses the theme.

Those caves relate to each other, creating lots of cracks. Vague rules, unconnected but continuous space, and settings that seems random but based on precise calculation...all these were added together to shape what we need.

本页：等候区。140~141 页：积木般的木质楼梯。
This page: Waiting area. pp.140-141: Block-like wooden stairs.

温馨绿洲——時玑皮肤护理工作室
Sweet oasis – SHIJI Studio

为了提高空间使用率，本案将原有空间进行了错层分布。

入户保留高耸的接待门厅，往内延伸相对低矮且具包裹性的私密的洽谈区和化妆区，二层和三层则为皮肤护理室。

室内采用了玛曼努墙面、水磨石地面，以及温暖的木材，块面功能清晰，感官专业、天然、舒适，在都市中心营造人们心中的一片现代绿洲。

To optimize space efficiency, we decided to distribute the taggered floor construction method.
As remaining the high gate for the reception hall, creating the business area and make-up area by laying out the lower inner ceiling, as well as the private hubs. The second and the third floor are used as skin care center.
We used Marmorino art paint for the wall inside the room, with terrazzo floor and wood decoration, all of which we believed would deliver clear professional and naturally comfortable surroundings, what's more, build a place like an oasis in the center of modern concrete forest, where people could relax and rest, also receiving cares.

接待区

Reception.

We are following the philosophy that
design is guided by our lives, concerning and getting
insight into what is truly people feeling for,
would always find us the way to demonstrate
precisely. Rather than seeing only one side of coin,
we ran deeper and even smaller degrees to discover
any potential priority. I would say, the predominant
features of this case are
innovation, restraint, fun, humor and fusion.
– Yang Wang

我们奉行以生活哲学指导设计，
关注和洞察人们的情感诉求，
并使其准确地加以呈现。
我们摆脱传统的片面思考，
善于从狭小的角度挖掘重点。
创新，克制，趣味，幽默，融合是我们的特点。
——汪洋

本页：VIP 区。右页：饰品陈列墙。

This page: VIP room. Opposite: Ornament shelf.

时尚之心——缇纱

The heart of fashion – Dejavu

缇纱是一家女装买手店，分一、二两层，整体空间由开放到私密逐步推进。

为展现这一优雅且充满活力的品牌，水磨石、黄铜、烤漆板、艺术涂料、绒布、纱帘，这些丰富的材质被组合使用。大面宽的街铺可视立面使用了高饱和度的蓝色与玫红色，以提高视觉冲击力，并增加粉色、灰色，以及四种颜色调和而成的粉紫色来柔和整个空间，形成愉悦、活泼、时尚的氛围。

Dejavu is a two-floors' women's dress boutique, the idea of design was to present the entire place a transition from open to private gradually. For revealing an elegant but also dynamic gene of the brand, a plenty of materials has been introduced, terrazzo, brass, lacquering board, art paint, velvet, and yarn curtain. People would feel overwhelmed seeing the wall of strong saturation blue and rose through the wide glass when walking by, along with the color of pink and grey, we added pinkish purple harmonized from four colors, to tone down the space in a delightful, lively and stylish atmosphere.

左页和本页：服装展示区及其装饰细部。

Opposite and this page: Clothing display area and its details.

TOWARDSFLORENCE
BIENNALE2019
CHINADESIGN
SELECTION

左页：展区入口。150 页和 151 页：建筑设计和时尚设计的展厅一隅。

Opposite: View of the entrance. p150 & p151: The corner view of the architectural design area and fashion design area.

黄金比例——佛罗伦萨国际（中国）设计双年展展厅

Golden ratio – The Florence International (China) Design Biennale

设计灵感源于达芬奇的人体黄金比例设计图，结合美术馆现有的场地特点，在空间中以嵌入矩形和圆形来达到动线的营造和展区的分割，在展馆中又强化了本次展览的主色调，从而创造出一种令人印象深刻的“佛双蓝”。

The design is inspired by the golden ratio design of human body from Leonardo da Vinci, combined with the existing site of the art gallery, and embedded rectangles and circles in the space to form a dynamic circulation and the division of exhibition areas. It also strengthens the main color of this exhibition, and thus creates an impressive “Blue”.

建筑设计
ARCHITECTURAL
DESIGN
陈展辉 ZHANHUI CHEN
陈政 CHEN ZHENG
方晓风 FANG XIAOFENG
ZHENG FU
杭州园林设计院
HANGZHOU LANDSCAPE ARCHITECTURE
DESIGN INSTITUTE CO.,LTD
九沐景观 JUMUJINGGUAN
郎水龙 LANG SHUILONG
李祥君、张飞
XIANGJUN LI, FEI ZHANG
王晨辉、吴锐
DONGLONG LIN, CHENHUI WANG, RUI WU
陆海 LU HAI
JIANJUN MAO
YUNHUA NIE
WENG JIANXIANG
SHUGANG YI
CHENFAN ZHANG
ZHENG DONGXIAN
ZHENG JIANG
ZHOU QI
ZHOU WEI
HUANG ZE , LUO XUAN
YUAN LIUJUN , XU YONGYONG
JIN JIE
林再国 LIN ZAIGUO
建筑设计
ARCHITECTURAL
DESIGN

FASHION
DESIGN

展厅之间的通道，“佛双蓝”和白色交替出现。
Passage connecting different exhibition areas.

展馆设计结合了不同作品的展示要求，以最佳观感来把握不同展区的特点，如时尚设计展区采用 T 台的展示手法，在圆形展厅内部做了足足十米长的白色展台，建筑设计展区用体块化的建造构成手法来表现，多媒体展区则运用纯黑色的布幔来达到吸音效果和营造暗环境，从而很好地呈现多媒体作品的视觉效果。

The design of the exhibition hall combines the display requirements of different works, with the best look and feeling to grasp the characteristics of the different exhibition areas. For example, the fashion design area has a full ten-meter-long white booth to show the performance of the T stage inside the round hall. The architectural design area is represented by a mass construction method, while the multimedia area uses pure black curtains to achieve sound absorption and dark environment, thus well presenting the visual effects of multimedia works.

左页：珠宝核心展区。
Opposite: Core display area of Jewelry.

神秘匣子——L&C World
Mysterious casket – L&C World

本案为一家搜罗全球奢侈品的买手店，商品涵盖服装、珠宝及创意产品。在 60 平方米的狭小空间内，根据原始结构的尺度和空间感，划分出服装、珠宝饰品、会客区三个功能区域。

It's a design case of multi-brand boutique (select shop), merchandise would cover clothes, jewelries and creative products. Of its' limited 60 square meters space, with fully respect the original structure and dimensions, we divided it into three functional areas felicitously, garments, accessories and lounge.

珠宝区。

Jewelry room.

设计灵感来自盛行于 17、18 世纪的满载财宝的海盗船。

The design inspiration comes from pirate ship prevailed in 17th and 18th century, whose image was full of treasure and curios.

服饰区。
Clothing room.

船上有人、有动物，停靠过海港，经历过巨浪，通过戏剧性的图案及颜色来表达情景和趣味性，试图让人产生代入感。

"Thinking about the exciting adventure, the ship carried sailors and animals, been through freak waves, eventually stopped some quiet harbors, and again repeat its journey. We selected dramatic wallpapers and colors to speak for this scene and fun, hoping to bring audiences back to the thrill scenario."

居住空间
Dwelling Space

左页：入口门厅。左边是垂直交通区，同时靠近地下室入口，业主回家后可以很方便地去往自己想去的地方。

Opposite: Hallway. Vertical transportation area is located on the left side, next to the entrance of the basement, where the homeowner can reach wherever he wants.

都市桃花源

Urban Peach Garden

林语堂说，“论文字，最要知味”，那么论设计呢？设计师觉得高品质不是材料的堆砌，更不是专业的炫技，而是表达出家的温度，以及主人的品味和个性。

该设计充分考虑本案业主的家庭情况和生活状态，在简约的空间表象下注入精致的细节设计。

As Yutang Lin says, “Taste is the most critical part in literary criticism.” So is designing. To achieve the so-called high quality is not to pile up materials or show off expertise, but to express the warmth of the home and to establish the taste and characteristics of the homeowner.

The designer takes the family and lifestyle of the homeowner into full consideration, and thus conducts exquisite designing within a simple space.

一层空间按照生活场景和动线来塑造。客厅以浅色为底，以咖啡色进行中和，以蓝色作为点缀。南侧全景落地窗将花园景致纳入室内，使得屋里大气亮堂，充满自然气息。餐厅分为中西两块：西餐厅为平常家用，临近厨房；而中餐厅则靠近庭院，更为宽敞，用于待客。这里不仅是一日三餐的功能区，也是业主与家人、朋友交流情感的场所。

The first floor plan is arranged according to the life scenes and circulation. The living room is decorated in light color, harmonic in brown and dotted blue. The floor-to-ceiling windows on the south side intorduce garden views into the interior, make the room full of light and nature. The Western dining room, next to the kitchen, is for daily use, while the more spacious Chinese dining room, closer to the garden, is used for hosting guests. It is not only a functional area for three meals a day, but also a place for owners to communicate with their families and friends.

左页：楼梯俯瞰。本页：平常家用的西餐厅。
Opposite: Downward view from the staircase. This page: The western dinning room for daily use.

从下沉式客厅望向入口。
Looking toward the hallway from the sunken living room.

本页：二楼休闲区装饰。右页：三楼起居厅边景。

This page: Close view of the relaxing space on the 2nd floor. Opposite: Corner of the livingroom on the 3rd floor.

DUMP RECIPES

本页：三楼起居厅。右页：三楼卫生间局部。
This page: The living room on the 3rd floor. Opposite: Corner of the bathroom on the 3rd floor.

业主夫妇的空间位于三楼，一条长长的过廊串联着起居室、衣帽间、卧室、书房，也串联起东南西三个阳台，不同的地面材质分隔着空间的内外。

The master bedroom is on the third floor. A long corridor connects living room, dressing room, bedrooms, study, as well as three balconies. The inner space and outer space of the room is seperated obviously by different floor paving materials.

漫步而行，内外皆是风景，而人的情感和心理亦在空间的移步换景中逐渐变化，从兴奋到安静，最终呈现人与空间的和谐相处。

Walking in the home, family members can feel scenery everywhere. Their feelings and emotions change with their pace, from exciting to calm, and eventually resonate with the space to reach harmony.

地下室空间。
Lobby in basement.

地下空间以三个天井作为切入点，以休闲运动为主题，围绕自然光照来布局。

设计师相信，设计创造的是体验，而不是空间或者陈设，就好比有品质的生活，享受的并不是物质本身，而是物质带来的美好感受。

The basement, which begins with three atriums, is designed with the theme of leisure and sports, and is planned with natural lights.

The architect believes that to design is to create experience, but not space or settings. It is the same concept that a quality life needs mental enjoyment rather than physical one.

地下室的茶空间。
The tea room in basement.

"Beauty is pleasure,
but not demand."
– Kahlil Gibran

"美不是一种需要,
而是一种欢乐。"
——纪伯伦

镜子中的餐厅一隅。
View of the dinning room in mirror.

琴·镜

Lyre & mirror

感性的惊喜往往依托于理性的排布而生，如同琴键上跳跃的音符，只有在音乐家的精心谱写下，才能演化成曼妙的音律。空间亦是如此，同样的户型，通过巧妙的设计，则可以在既定的规范之中，营造独一无二的私人领地。

空间的个性在一定程度上呈现着业主的气质。简约、时尚、艺术便是欢乐海岸这座私宅所呈现的气息。中性色的空间既有黑白灰的冷静，也有木森系的温润；由布艺、皮质与金属搭配的家居，雅致中透露着时尚；伯恩斯坦钢琴与现代主义的画作相得益彰，丰盈着家里的艺术气息。

Surprises often come out based on rational arrangement, just like music notes change to be great rhythms only under finest composing. So is designing. Even a common home can be transformed as a unique personal space under certain conditions via exquisite design process.

The attribute of the space reflects the character of user to some extent. The atmosphere of this project appears to be simple, stylish and artistic. Neutral colors contain the calmness from black, white, and grey, as well as warmth from natural colors. Materials for furniture are mainly cloth, leather, and metal, establishing elegance and fashion. The artistic aura fills throughout the home with Bernstein piano and Modernist paintings.

客厅，大面积的开窗让室内显得明亮开阔。
The living room with a big window makes the interior much brighter.

晨起，带一杯咖啡来到窗边，悠闲地坐下，阳光洒在茶几上，也洒在窗外的奉化江上，江面的微光和远处的白云唤起人一天的好心情。客餐厅是敞开式的，多来几个朋友也不会显得拥挤。兴致来了，或许就打开琴盖弹上一曲。琴音悦耳，香茶醉人，嘉朋满座，欢乐自然会在心中流淌。

In the morning, the homeowner can sit at the window with a cup of coffee, enjoy sunshine, and watch Fenghua River outside flashing even more natural light. A good mood is therefore brought up with all the scenery for the whole day. There is an open living-dining room that can host parties for friends. This space is widely opened for different purposes. When we play the piano, pleasure arises with great music, tea, and friends.

钢琴角。

Piano corner.

镜中花月，盈盈有诗意。

A round mirror is very well located in the dinning hall, as if it could reflect pleasant stories from a parallel world.

镜中客厅一隅。
View of the living room in mirror.

家有时就像生活的镜子，照见生活的日常，也反映诗意的芳华。而设计的魔力就在于，唤醒人对美好的渴望，让日常成诗。

Home is a mirror of life, reflecting both routing and poetic. The magical power of designing is to activate people's desire of beauty, and to bring more poetic feelings.

左页：客厅壁炉区，温暖的火光，也给人带来心理上的温暖。182~183 页：客厅。

Opposite: View of the fireplace in the living room, the warm firelight brings warmth to people psychologically. pp.182-183: View of the living room seen towards the dinning room.

泊景秋月白

Bojing moon white

这是一个 1050 平方米的私宅。从设计到落成历时 3 年多。建筑本身外立面简洁，呈退台式结构，有入户庭院、下沉式庭院、天井，节奏感很好。设计顺建筑之势，以自然性和人文性为基调。

This is a villa house of 1050 m^2, which takes three years to build. The facades, with a set-back on one side, are clean and neat. There is an entrance courtyard, a sunken courtyard, and an atrium in the house. The rhythm of the house is well tuned, following the building's own melody, with nature and humanity as the main tone.

采用轻软装化的设计，更多地关注空间在结构上的层次关系和丰富的体验感。

没有电视的客厅，留白，使“人”成为焦点，从而拉近家庭成员之间以及主宾之间的关系，也是对我们热衷于社交软件沟通的一种逆行反思。

Interior design is lightened to give focus to relationship of structural layers and rich experience.

A living room without TV, also known as leaving the blank, focuses on “people”, draws family members closer, as well as eliminates the distance between hosts and guests. The design is also a contradiction of relying on social networking applications.

左页：扶手细节。本页：从客厅上去就是电梯。
Opposite: Details of the handrail. This page: The elevator is next to the living room.

地下室家庭活动区。
Family activity room in basement.

地下功能厅作为家庭活动区，家庭成员在此停留的时间会更多一些，其设计相对更温馨一些，在陈设上也更饱满一些。地下室的阁楼设计成了小孩的独立空间，对外可俯瞰小天井，对内可通过一个条形窗与楼下的父母互望，在大人和小孩之间建立起微妙的视觉和心理联系，空间也因此交融起来。

The multi-function hall in basement is a family activity room; therefore, there are more contents here when it is assumed that the family will spend more time here. The top portion of the basement is designed for children where they can look down on the atrium, and keep eye contacts with their parents through a bar-shaped window. Thus, the spaces interact with each other visually and psychologically.

左页：地下室一景。本页：楼梯下的小空间。
Opposite: View of the basement floor. This page: The sweet space under the stairs.

为了减少楼梯的高耸感，在其下方做了两个台阶做铺垫，并延伸到楼梯下方，形成一个小空间，装上灯带后，产生了光影的交流，小空间呈现出私密而温馨的状态，成为小宝宝最喜欢的地方。

The stair shows extra height than it actually has. To eliminate this feeling, two more steps are set under the stair, eventually forming a small space. With strip lights assembled, the space is granted more privacy and warmth, becoming a favorite place for the baby.

地下室水吧。
The water bar in the basement.

合适的设计是一种克制，而美的设计体现了人文性、自然性和艺术性，不是仅仅用装饰就可以概括。在这个项目里，采用了很多留白，让人有更多的思绪停留，在空间当中感受人与人、人与自然的对话。

The architect believes that the right design comes with restriction. He also believes that aesthetics design is the combination of humanity, nature and art, much more than decoration. In this project, there is lots of blank space, letting thoughts to stay, to communicate with oneself, and to realize the dialogue between human and human, and between human and nature.

人和人的交流不一定仅发生在同一个空间里，它还可以穿越楼层和墙体的隔离。

The communication between people does not have to happen in the same place; instead, it can also cut through walls and floors.

本页：由原来的书房改成的餐厅。196~197 页：客厅和过道。

This page: The dinning room which was renovated from a study. pp.196-197: The living room and corridor.

宁静的自由

Liberty in peace

哲人罗素说：“在安静的气氛中，才能产生真正的人生乐趣。”设计以重组布局的方式重新定义空间的功能和尺度，放大空间景观和面积的优势，去掉冗余的功能，优化空间的动线，并将多余的柱体和设施遮挡，使之消融于立面的体块构成与美陈之中，让安静、优雅和包容成为空间的三个关键词。

Great philosopher Russell once said, “True happiness can only be found in a quiet life.” By regrouping the layout, the functions and scales are redefined, the advantage of views and size of the flat is enlarged, the redundant parts are eliminated, the circulationt is optimized, and the unused pillars and settings are hidden. The space is thus expressing the main thesis of quietness, elegance and inclusiveness.

客厅壁炉空间。

The fireplace in the living room.

居者在设计的引导下走向不同的空间，从书房可以隐约看到客厅的壁炉一角，这种相互的透视让空间变得更有意趣。夜幕降临，家的氛围在智能灯光的呼吸中被慢慢唤醒。在这宁静的场域之中，光线轻柔地漫开，触摸着素雅的墙面，划过大理石纹的地面，将家中一切的美好映入你的眼帘。当你漫步时，空间后退为背景；当你停留时，周边是你的舞台，于是回家成为一种幸福的体会。

Homeowner is guided to different rooms by the design. The fireplace in the living room can be seen from the study, establishing a clear vision, making the space more interesting.

When night arrives, smart lights are awakened, creating an atmosphere of warm home. In this calmness, lights are casted out gently, touching walls, gliding along marble floors, and eventually bringing all the goodness to you. When you walk in the home, you can feel the background fading; when you stop your step, you can feel the stage surrounding. Home sweet home.

左页：茶空间；本页：小憩空间。
Opposite: Space for tea; This page: Space for leisure.

正如陈果所说，当一个人遵从自己的内心，明确了有所为有所不为的精神自律，放弃一些世俗的浓艳芬华，恰是他选择和维护了他心向往之的那种自由。

As Chen Guo said, “Only when one follows his or her heart, realizes the self-discipline of dos and don’ts, and gets rid of secular thoughts, can he or she approach beloved true liberty.”

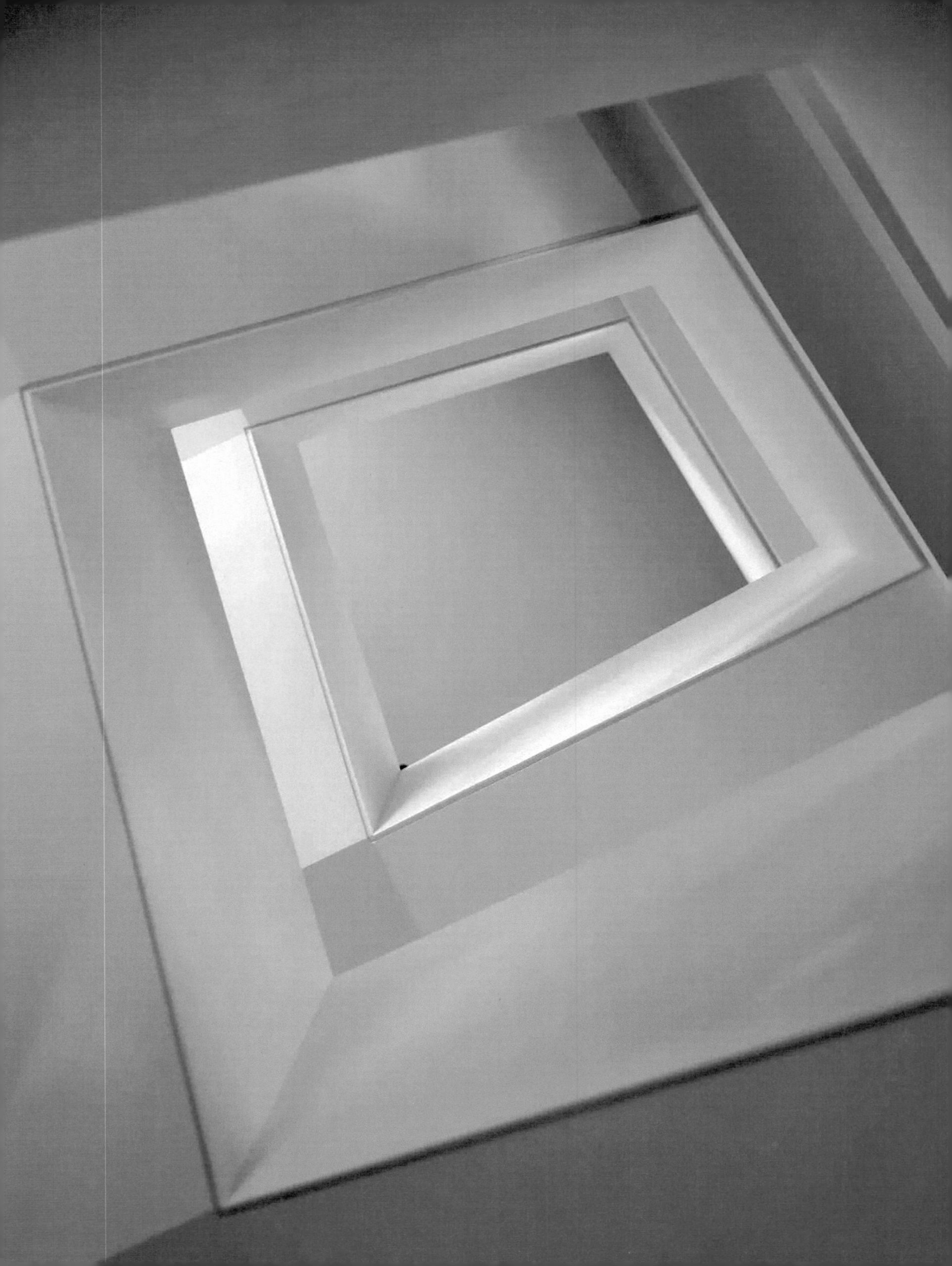

左页：不规则的异形楼梯。本页：楼梯空间。

Opposite: Looking up the irregular-shaped stairs. This page: View seen from the stairs.

共融共生

Harmonious symbiosis

在设计师看来，美是关于生活与自然的自由艺术，家的设计则是对人与生活的解读和理解，是人与空间共融共生的形态。不规则的异形楼梯像一个艺术装置从地下室上升至三层，诠释着刚柔并济的艺术形态，兼顾建筑与雕塑的艺术之美。

From the architect's view, beauty is a free art form about life and nature, while the design of a home is an understanding of people and life, and it is the symbiotic form of people and space. Rooted in the basement and ascended to the third floor, the irregular-shaped and art-like stairs demonstrate limber and solid forms of art, which contains the beauty of both architecture and sculpture.

左页：客厅。本页：餐厅。
This page: The living room. Opposite: The dinning room.

项目将极致设计美学与当代艺术融合，朗阔的空间透露着一种安静的力量。大面积的落地玻璃窗与户外花园相通，延伸视觉的艺术之感。

Infused with aesthetics and contemporary art, this project reveals quiet power in the grand air. The artistic atmosphere extends from indoor area to courtyard garden through large floor-to-ceiling windows.

呈现一个自然、舒适、现代化的艺术居家空间。

A natural, pleasant, modern and artistic home is hereby well presented.

茶室。

The tea room.

To design is not simply to invent, but to pursuit the
nature and reality. A design looks like a journey and a
dialogue between the mind and the nature.
The pure sky, crystal blue ocean, vast forest...
all of these can bring me amazement and feelings.
– Suping Zhuo

设计不是设计事物本身，
而是人对于自然与本真的向往。
每一次的设计好像是一场旅行，
都是心灵与自然的对话。
纯净的天空、蔚蓝的海洋、广袤的森林……
都能带给我不同的惊喜与感动。
——卓稣萍

左页：地下室庭院。
Opposite: Courtyard in the basement.

慵懒的家
Home with lazy style

业主是一位资深的高端家居品牌经营者，与设计师是多年的好友，在设计这套房子的时候给予了设计师绝对的信任和放权。当然，这一切都基于相互了解。

设计更改了入户动线，从院门进入走一条折线，将入户门开到了房子的中部位置，使得入户时有曲径通幽、豁然开朗的仪式感。更重要的是，原先“凹”字形的细腰结构，在改动之后，完成了对细腰部分的结构弥补，二楼、三楼变成了“口”字结构，空间不但更为方正，而且实际利用面积也变得更为充裕。

The homeowner runs a high-level home supplies enterprise for several years. The architect is a friend of hers for decades, so he was granted full authority to design this home, given the fact that he know her lifestyle very well.

There were three main flaws on door-opening position because of the linear-like floor plan. Therefore, the entrance circulation is modified as a key feature for the new home. Now the front door is around the middle of the whole house, with a polyline-like path from entrance gate, which brings a better back-to-home experience, and a better solution for storage space and rain shelter. What is more, the structural gap of the original house is filled up and the house looks more like a block, with practical areas increasing largely.

从入口玄关望向餐厅。

Dinning room view from hallway.

房子的入口改变之后，入户有了一个漂亮的玄关，左边是客厅和花园，右边是餐厅和厨房。无论动线、格局还是空间的尺度都非常舒服，而楼梯就在入户门的右手边。可以说，入户所处的动线核心位置，使得平面交通和垂直交通都十分方便，视线和行动都没有什么阻碍。

Thus, a decent hallway is presented behind the front door. There is a living room and courtyard garden on the left side, and dining room, kitchen and staircases on the right side. The circulation, allocation, and scale of the space are all enjoyable, bringing a great convenience on both horizontal and vertical traveling. There is no obstacle to block vision or movement either.

The core for modern space designing
is not vision or style, but the warmth of home.
– Qifeng Zhang

现代空间的设计核心，
并不是视觉和风格。而是家的温度。
——张奇峰

安静的客厅一角。
A corner of the lving room.

客厅的设计让人眼前一亮，这里的视觉呈现十分当代和前卫，写意空间的客厅家具，搭配 vertigo 灯（草帽灯），以及巴瓦那女郎叼着雪茄的酷照，呈现一种独立女性的自信和洒脱。

The living room is designed to amaze. The attribute of the space here is modern and stylish. Ligne Roset furniture was put here, with Vertigo lamp and cool picture of a Bhavana lady who is smoking a cigar, demonstrating confidence and freedom of the female.

俯视“懒人空间”。
Looking down on the lazy area.

女主人爱养猫，设计师常常打趣她是属猫的。因为她的个性里蕴含着猫的敏感和慵懒，所以，他应用了懒人视觉的手法来打造这块区域。什么是“懒人视觉”呢？就是我们常常说的“葛优躺”的视觉中心，在这里整个视线是被拉低的，这里挂的灯离地面只有 90 厘米，画的中心点也只有 1.2 米高（我们正常的视觉中心应该是高 1.5 米左右），再加上趴地而坐的写意空间的“哈巴狗”椅，懒人空间就打造完毕。

The homeowner loves cats, and she is somehow sensitive and lazy just like cats in the designer's eyes. As a result, he applies a concept – “lazy vision” in the living room. So what exactly is “lazy vision”? Basically speaking, the center of vision is lower, as if you are laying on sofa. Everything here is lowered: lights are 90 cm away from floor, and paintings on the wall are 120 cm away from floor, of which the common number is 150 cm. All these settings plus a Ligne Roset “pug” chair complete this lazy space.

客厅的懒人视觉中心。

The "lazy vision" area in the living room.

有鉴于彼此之间的了解，设计师可以想象到她在这个空间中的状态，想象她如何进家门，如何坐下，如何品茶阅读，如何与她的爱猫嬉戏，她的视线在什么位置，友人的书画如何和空间和谐，灯光如何配置……所以这个家的很多物件的高度以及器材的位置其实是专为她而设计的，让她在家里保持轻松的状态，看到家里最美的一面。

Because he knows the homeowner very well, he can picture how she lives in this house, such as how she enters the front door, how she sits, how she drinks tea and reads, how she plays with her cats, how she casts her sights…he can also image where to put her friends' paintings and how to set lights…as a result, most items and equipment are designed solely for her, and at her ease. So that she can stay relaxed and keep comfortable at home, and see the most beautiful view of her house.

左页：衣帽间。本页：楼梯过道。

Opposite: The dressing room. This page: View of the staircase.

生活的许多趣味，
就在于
和自己喜欢的东西待在一块，
然后家就可以变得很有温度，
那是一种暖到心里的感觉。

Fun of the life often comes with
staying with things one love.
A home is then a sweet home,
warming one's heart.

Bringing Nature
ROOMS FOR LIVING

左页：客厅的壁炉空间。

Opposite: View of the fireplace in the living room.

空灵维度

Ethereal dimension

业主向往的生活，纯粹干净，浸润着艺术的高雅与质朴。她希望家里的艺术氛围盈盈袅袅，以亲近的姿态融入生活场景，而不是艺术品的简单堆砌或过度填塞。

设计师始终关注这一诉求，拿捏尺度，以适当的留白，将艺术想象力还给生活。

The life that the homeowner wishes for is clean and pure, with artistic elegance and simpleness. She loves to fill her home with art in a natural manner, instead of simply piling up a bunch of art pieces.

The architect keeps that in mind, seeks the appropriate scale, and gives back life some artistic imagination with proper blank in designing.

整体以开阔的布局、明朗的线条为主，营造出惬意的安适感；以珍珠白与木作棕为主色调，借助色彩在感官和精神内涵层面引发美的通感。白创造了光洁简约的空间表面，强化了人对空间框架和秩序的感知，再加入木作、大理石材家具，透过纹路、肌理散发出自然朴拙的侘寂感。空间里有自然的丰盈，有时间的光泽，有生活的诗意。

The layout is relatively loose, together with straight lines, to establish enjoyment and comfort. Pearl white and wooden brown are main colors. White color creates smooth and simple surfaces, enhancing the perception to the frames and order of the space. Wooden and marble furniture is used to creat a sense of nature and calmness with unique texture.

The architect uses colors to bring beauty from vision and mental perspective, creating a true sense of belonging within every inch of the home.

左页：客厅。本页：餐厅边柜，抬头可见艺术画作。

Opposite: The living room. This page: The dinning room with a painting on the wall.

餐厅全景。

General view of the dinning room.

地下室保留了 4.8 米的层高，用壁炉、挂画和天井，营造了一个会客的空间与艺术长廊。

A living room like an art gallery is in the basement with height of 4.8 m, ornamented with fireplace, hanging paintings and an atrium.

设计无痕
Traceless design

整套别墅建筑占地仅 90 平方米，但套内实用面积达到了 360 平方米 。设计师打开部分墙体，扩大窗户面积，以开放式的空间设计，给人以开阔的美景视野和舒适的居住体验。简练的线条勾勒搭配通透的设计，让整个家充满阳光般的生活格调。

The site area of the villa is only 90 square meters while its floor area is as large as 360 square meters. To fit the natural environment, the architect removed part of the wall to enlarge the area of windows, and thus brought an open view and enjoyable living experience with this open-ended design method.

Simple lines and crystal clear design make the home full of life of sunshine.

本页：壁炉空间。右页：从壁炉望向门口。
This page: View of the fireplace. Opposite: View toward the hallway from the fireplace.

通透的空间设计，不仅利于通风与采光，让整栋房子的每个空间都能与大自然完美融合，而且增加了不同区域的互动与交流。源自意大利的顶级现代奢华家具、德国的橱柜五金，以及当代艺术挂画，家里的每一件器物都融入了独特的思想和理念，组成了和谐时尚的人居艺术空间，透露出一种越简洁、越精致的设计气质。

Simple and clear design not only makes the building easier for ventilation and capturing natural lights, mixing every room with the nature, but also adds interaction and communication among various areas. Every furniture and equipment in the home, such as top tier luxury furniture from Italy, cabinet and hardware from Germany, and modern artwork drawing, is infused with unique concept, forming harmonious and artistic space, demonstrating the atmosphere of "the simpler, the more delicate".

左页：顶楼阳光房。本页：楼梯墙面的光与影。
Opposite: The sunlight room on the top floor. This page: Light and shadow on the wall side of the stairs .

顶楼，作为自然光线最为丰富的地方，设计师把这里留给了儿子，卧室和阳光房的组合给了他自由、光明且充满生机的空间，大面积的落地推拉门可以自由开合，让卧室和阳光房完美融合，为家融入了更多的自然属性。

简约，不一定要冷淡，而是让自己有更多的精力对待自己喜欢的事情。它可以很有趣，比如，没有任何装饰的楼梯墙面在灯光打开之后，那种光影的变化让整个垂直动线变得丰富而活泼；仅仅通过沙发靠背位置的调整，一个客厅就可以完成不同类型会客模式的切换；在平整的卧室墙面上轻轻一推，一道通往浴室的门就被平滑地打开了……这些充满趣味的空间变化就是简约设计中蕴含的灵魂和思想 。

As the space is filled with most natural light, the top floor is designed for his son. The combination of bedroom and sunlight room gives him a free, bright and vigorous room. The floor-to-ceiling door of the room makes the mixture of bedroom and sunlight room perfect, infusing more natural elements into the home.

Simple does not necessarily mean coldness, because it aims to eliminate clutter and pay more attention to things one likes, but not to take off the fun of life. We see the rich and lively variation of light and shadow of vertical lines on the clean wall with no decoration when the lights were turned on; we see the living room shifting into different meeting areas by means of simply adjusting the back cushion of the sofa; we also see a bathroom door opening smoothly after a gentle push on the smooth bedroom wall…all the fun space varieties are the soul of simple design.

The true design is not to amaze someone via superficial decoration, but to achieve family harmony through scenario designing.
– Qifeng Zhang

真正的设计不是通过表面的装饰让你惊艳，
而是通过场景的设计，让家庭变得更加和谐。
——张奇峰

厨房设计是一个非常重要的环节。在这个空间中，设计师设计了这样一个场景：当女主人在厨房做菜的时候，男主人就坐在岛台旁边，陪她聊天。

所以设计师将原来与厨房相邻的卫生间功能下放到了楼道夹层中，扩大了整个厨房的面积，并通过隔而不断的吧台和玻璃移门，使厨房和餐厅之间形成了有效的互动，为家庭生活添加更多的可能性，让家人们有了更多接触的时间。

Kitchen is a significant part. In this space, the architect sketches such a picture: when the hostess is cooking the meal, the host sits at the central island and chats with the hostess.

Therefore, the designer shifts the bathroom that originally next to the kitchen down to the mezzanine, enlarging the kitchen area. The designer also sets some bar counters and glass sliding doors to make interactivity between kitchen and dining room, add more life possibilities, and create more family time.

从夹层看厨房上空。
View of kitchen from the mezzanine.

地下室的书桌。
Study desk in the basement.

I prefer two extreme design concepts. One is extremely far away from life – simple, silent, rational, distant. While approaching the limit of this extreme end, this space appears to have better quality, atmosphere, and aura. The other is extremely close to life – every brick and every item used can all be traced back to real life. Books can be reached within inches, and relaxation places can be found around the user. With sunlight and plants surrounding, the feeling of beautiful life comes into users' heart.

– Qifeng Zhang

我喜欢两个极端的设计，
一个极端是出世的设计，
至简，至寂，至理，与人有距离感，
但显得更高级，大气，更有意境；
另一个极端是入世的设计，
一砖一瓦，一物一什都是生活的痕迹，
随手能翻到书，随地就有小憩的空间，
阳光映入，绿植绕身，最美人间烟火味！

——张奇峰

本页：房间一隅。238~239 页：客厅，背景选用了一幅铁凝铁艺作品，用于平衡各方面的元素，让家呈现一个比较和谐的东方主题。

This page: A corner of bedroom. pp.238-239: The living room, with an art piece made of ironis on the wall, to balance all the elements and express harmonious oriental theme.

画境之家

Home with artistic mood

很多人以为，设计师的家应该是天马行空的，是设计师随心所欲的作品，是一个极端或另一个极端，但设计师郑钢说：未必！家的设计不是艺术家的独白，它不是一个人的事情，而是一家人共同意愿的结晶。对于他们而言，设计自己的家和设计客户的家，本质上是一样的，都要尊重家庭成员的意见，都要受到客观条件的制约。

Some people believe that designing is supposed to be dramatic, going to either one extreme end or the other one. But Mr. Gang Zheng gives an opposite perspective. He claims that designing a home is not monologue of the designer, but the effort of a whole family. Designing his own home is the same as doing that for clients, which needs intergrating the opinions from family members and is restricted by various limitations.

本页：厨房一角。右页：餐厅。242~243 页：变身为书房的餐厅。

This page: Kitchen. Opposite: The dinning room. pp.242-243: A study transformed from the dinning room.

由于经常在家里吃饭，厨房空间和餐厅的设计就尤为重要。餐厅可以随时进行角色转换，可以变为书房，也可以变成茶室。相对于视觉上的愉悦，设计更大的价值是给生活在家里的人带来生活品质的提高和精神的愉悦。

Kitchen and dining room are the most critical spaces because the family members dine at home a lot. The dining room is flexible. It can be transformed into a study or a tearoom whenever the homeowner wants. The most valuable outcome of designing a home is to bring high quality life and enjoyable atmosphere rather than vision pleasure.

变身为茶室的餐厅。

The tearoom transformed from the dinning room.

There is no perfection in designing.
Architects are on the way to achieving
the ideal design, where they try their best
to catch every inspiration and
take one step closer to their dreams.
– Gang Zheng

世上没有完美的设计，
设计师只是在追寻理想设计的道路上，
抓住每一个灵感机遇，
让自己离梦想再近一点。
——郑钢

项目信息
Project information

文化休闲 Cultrural Space

项目名称 | Name: 望春堂 Wangchun Church (pp.2-9)
项目地址 | Address: 浙江宁波 Ningbo, Zhejiang
建筑面积 | Area: 620 m^2
完成日期 | Compeletion: 2020.7
主案设计 | Chief Designer: 卓稣萍 Suping Zhuo
参与设计 | Team Members: 卓永旭、王旭辉 Yongxu Zhuo, Xuhui Wang
撰文 | Writer: 卓稣萍 Suping Zhuo
摄影 | Photographer: 刘鹰 Ying Liu

项目名称 | Name: 申山乡宿一号别院 #1 Villa of Shenshan Rural Inn (pp.10-21)
项目地址 | Address: 浙江衢州 Quzhou, Zhejiang
建筑面积 | Area: 780 m^2
完成日期 | Completion: 2018.7
主案设计 | Chief Designer: 李财赋 Caifu Li
软装美陈 | Soft Decor: 胡荣 Rong Hu
参与设计 | Team Members: 赵铁武、王军政、胡荣海 Tiewu Zhao, Junzheng Wang, Ronghai Hu
撰文 | Writer: 叶建荣、袁玥 Jianrong Ye, Yue Yuan
摄影 | Photographer: 刘鹰、阿坡 Ying Liu, A Po

项目名称 | Name: 曦所 Xi Club (pp.22-31)
项目地址 | Address: 浙江宁波 Ningbo, Zhejiang
建筑面积 | Area: 110 m^2
完成日期 | Completion: 2018.5
主案设计 | Chief Designer: 胡梁峰 Liangfeng Hu
撰文 | Writer: 叶建荣 Jianrong Ye
摄影 | Photographer: 刘鹰 Ying Liu

项目名称 | Name: 大步里院 Dabu Courtyard (pp.32-39)
项目地址 | Address: 浙江宁波 Ningbo, Zhejiang
建筑面积 | Area: 2000 m^2
完成日期 | Completion: 2016.8
主案设计 | Chief Designer: 李财赋 Caifu Li
软装美陈 | Soft Decor: 胡荣 Rong Hu
参与设计 | Team Members: 赵铁武、胡荣海、潘宏建 Tiewu Zhao, Ronghai Hu, Hongjian Pan
撰文 | Writer: 叶建荣 Jianrong Ye
摄影 | Photographer: 刘鹰 Ying Liu

项目名称 | Name: 慧舍 Hui House (pp.40-49)
项目地址 | Address: 浙江宁波 Ningbo, Zhejiang
建筑面积 | Area: 120 m^2
完成日期 | Completion: 2017.9
主案设计 | Chief Designer: 潘高峰 Gaofeng Pan
参与设计 | Team Members: 高慧琼、陈燕南 Huiqiong Gao, Yannan Chen
撰文 | Writer: 潘高峰 Gaofeng Pan
摄影 | Photographer: 刘鹰 Ying Liu

项目名称 | Name: 楠山南 South of Mount Nan (pp.50-57)
项目地址 | Address: 浙江宁海 Ninghai, Zhejiang
建筑面积 | Area: 800 m^2
完成日期 | Compeletion: 2017.8
主案设计 | Chief Designer: 刘志明 Zhiming Liu
参与设计 | Team Members: 周敏刚、任军杰 Mingang Zhou, Junjie Ren
摄影 | Photographer: 刘志明 Zhiming Liu

项目名称 | Name: 杭州振邦律师事务所办公室 Office of Zhenbang Law Firm in Hangzhou (pp.60-67)
项目地址 | Address: 浙江杭州 Hangzhou, Zhejiang
建筑面积 | Area: 1100 m^2
完成日期 | Compeletion: 2020.8
主案设计 | Chief Designer: 卓稣萍 Suping Zhuo
参与设计 | Team Members: 卓永旭、覃小莉、王旭辉、温宇辰 Yongxu Zhuo, Xiaoli Tan, Xuhui Wang, Yuchen Wen
撰文 | Writer: 叶建荣 Jianrong Ye
摄影 | Photographer: 金选民 Xuanmin Jin

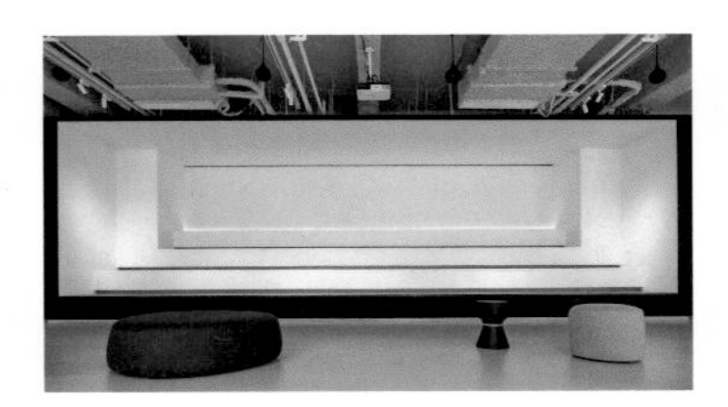

项目名称 | Name: 玉米之家 Office of Corn Design (pp.68-77)
项目地址 | Address: 浙江宁波 Ningbo, Zhejiang
建筑面积 | Area: 1300 m^2
完成日期 | Completion: 2019.12
主案设计 | Chief Designer: 郑钢 Gang Zheng
参与设计 | Team Members: 蒋哲淼、顾晓斌、朱朝华、陶鑫婷、孙红平 Zhemiao Jiang, Xiaobin Gu, Zhaohua Zhu, Xinting Tao, Hongping Sun
撰文 | Writer: 许露琪 Luqi Xu
摄影 | Photographer: 朴言、张学泉 Yan Pu, Xuequan Zhang

项目名称 | Name: 浙江欧硕律师事务所办公室 Office of Zhengjiang Oushuo Law Firm (pp.78-89)
项目地址 | Address: 浙江宁波 Ningbo, Zhejiang
建筑面积 | Area: 2000 m^2
完成日期 | Compeletion: 2019.12
主案设计 | Chief Designer: 卓稣萍 Suping Zhuo
参与设计 | Team Members: 卓永旭、覃小莉、王旭辉、温宇辰 Yongxu Zhuo, Xiaoli Tan, Xuhui Wang, Yuchen Wen
撰文 | Writer: 叶建荣 Jianrong Ye
摄影 | Photographer: 刘鹰 Ying Liu

项目名称 | Name: 宁波旷世智源办公室 KWUNG'S Ningbo Office (pp.90-97)
项目地址 | Address: 浙江宁波 Ningbo, Zhejiang
建筑面积 | Area: 2900 m^2
完成日期 | Compeletion: 2018.7
主案设计 | Chief Designer: 李财赋 Caifu Li
软装美陈 | Soft Decor: 胡荣 Rong Hu
参与设计 | Team Members: 赵铁武、胡荣海、王军政 Tiewu Zhao, Ronghai Hu, Junzheng Wang
撰文 | Writer: 叶建荣 Jianrong Ye
摄影 | Photographer: 刘鹰 Ying Liu

项目名称 | Name: 可瑞舒适家杭州总部 Co-Real HOME SMART in Hangzhou (pp.98-109)
项目地址 | Address: 浙江杭州 Hangzhou, Zhejiang
建筑面积 | Area: 1800 m^2
完成日期 | Completion: 2019.8
主案设计 | Chief Designer: 毛赟 Yun Mao
参与设计 | Team Members: 胡雅婷 Yating Hu
撰文 | Writer: 叶建荣 Jianrong Ye
摄影 | Photographer: 朴言 Yan Pu

项目名称 | Name: 博洋前洋 26 联合办公 Beyond Qianyang 26 Joint Office (pp.110-119)
项目地址 | Address: 浙江宁波 Ningbo, Zhejiang
建筑面积 | Area: 4000 m^2
完成日期 | Completion: 2019.4
主案设计 | Chief Designer: 毛赟、刘宏裕 Yun Mao, Hongyu Liu
参与设计 | Team Members: 胡雅婷 Yating Hu
撰文 | Writer: 陈贝贝 Beibei Chen
摄影 | Photographer: 刘鹰 Ying Liu

项目名称 | Name: 集艺办公室 Jiyi Office (pp.120-125)
项目地址 | Address: 浙江宁波 Ningbo, Zhejiang
建筑面积 | Area: 680 m^2
完成日期 | Compeletion: 2017.10
主案设计 | Chief Designer: 刘志明 Zhiming Liu
参与设计 | Team Members: 周敏刚、任军杰、傅行 Mingang Zhou, Junjie Ren, Xing Fu
撰文 | Writer: 刘志明 Zhiming Liu
摄影 | Photographer: 朴言 Yan Pu

项目名称 | Name: 浩然办公室 Haoran Office (pp.126-129)
项目地址 | Address: 浙江宁波 Ningbo, Zhejiang
建筑面积 | Area: 250 m²
完成日期 | Compeletion: 2017.10
主案设计 | Chief Designer: 潘高峰 Gaofeng Pan
参与设计 | Team Members: 高慧琼、汤伟海 Huiqiong Gao, Weihai Tang
撰文 | Writer: 潘高峰 Gaofeng Pan
摄影 | Photographer: 刘鹰 Ying Liu

商铺展示 Shop & Exhibition Space

项目名称 | Name: 烤古烧烤 Kaogu BBQ (pp.132-137)
项目地址 | Address: 浙江宁波 Ningbo, Zhejiang
建筑面积 | Area: 200 m²
完成日期 | Completion: 2014.6
主案设计 | Chief Designer: 毛赟 Yun Mao
参与设计 | Team Members: 胡雅婷 Yating Hu
撰文 | Writer: 毛赟 Yun Mao
摄影 | Photographer: 刘鹰 Ying Liu

项目名称 | Name: 時玑皮肤护理工作室 SHIJI Studio (pp.138-143)
项目地址 | Address: 浙江宁波 Ningbo, Zhejiang
建筑面积 | Area: 180 m²
完成日期 | Completion: 2020.1
主案设计 | Chief Designer: 汪洋 Yang Wang
参与设计 | Team Members: 王力宇 Liyu Wang
视觉表现 | Design Visualiser: 方露 Lu Fang
撰文 | Writer: 汪洋 Yang Wang

项目名称 | Name: 缇纱 Dejavu (pp.144-147)
项目地址 | Address: 浙江宁波 Ningbo, Zhejiang
建筑面积 | Area: 350 m²
完成日期 | Completion: 2018.11
主案设计 | Chief Designer: 汪洋 Yang Wang
撰文 | Writer: 汪洋 Yang Wang
摄影 | Photographer: 刘鹰 Ying Liu

项目名称 | Name: 佛罗伦萨国际（中国）设计双年展展厅 The Florence International (China) Design Biennale (pp.148-153)
项目地址 | Address: 浙江宁波 Ningbo, Zhejiang
建筑面积 | Area: 1280 m²
完成日期 | Completion: 2019.7
主案设计 | Chief Designer: 毛赟、胡雅婷 Yun Mao, Yating Hu
参与设计 | Team Members: 郎帆 Fan Lang
平面设计 | Visual Designer: 左右设计、启仓设计、形而上设计、何谓设计、为什么设计 Lright Design, Qicang Design, Sens Design, Hewei Design, Why Design
撰文 | Writer: 毛赟 Yun Mao
摄影 | Photographer: 一川黑水 Black Water Team

项目名称 | Name:L&C World L&C World (pp.154-157)
项目地址 | Address: 浙江宁波 Ningbo, Zhejiang
建筑面积 | Area: 60 m²
完成日期 | Completion: 2015.8
主案设计 | Chief Designer: 汪洋 Yang Wang
参与设计 | Team Members: 王力栋 Lidong Wang
撰文 | Writer: 汪洋 Yang Wang
摄影 | Photographer: 汪洋 Yang Wang

居住空间 Dwelling Space

项目名称 | Name: 都市桃花源 Urban Peach Garden (pp.160-173)
项目地址 | Address: 浙江宁波 Ningbo, Zhejiang
建筑面积 | Area: 1800 m²
完成日期 | Completion: 2019.12
主案设计 | Chief Designer: 张奇峰 Qifeng Zhang
软装美陈 | Soft Decor: 袁霞 Xia Yuan
参与设计 | Team Members: 沈璐、梁燕华、叶斌 Lu Shen, Yanhua Liang, Bin Ye
撰文 | Writer: 叶建荣 Jianrong Ye
摄影 | Photographer: 刘鹰 Ying Liu

项目名称 | Name: 琴 • 镜 Lyre & mirror (pp.174-179)
项目地址 | Address: 浙江宁波 Ningbo, Zhejiang
建筑面积 | Area: 208 m^2
完成日期 | Compeletion: 2019.12
主案设计 | Chief Designer: 卓稣萍 Suping Zhuo
参与设计 | Team Members: 徐群莹、任思玥 Qunying Xu, Siyue Ren
撰文 | Writer: 叶建荣 Jianrong Ye
摄影 | Photographer: 刘鹰 Ying Liu

项目名称 | Name: 泊景秋月白 Bojing moon white (pp.180-193)
项目地址 | Address: 浙江宁波 Ningbo, Zhejiang
建筑面积 | Area: 1050 m^2
完成日期 | Completion: 2020.4
主案设计 | Chief Designer: 李财赋 Caifu Li
软装美陈 | Soft Decor: 胡荣 Rong Hu
参与设计 | Team Members: 潘叶航 Yehang Pan
撰文 | Writer: 叶建荣、李财赋 Jianrong Ye, Caifu Li
摄影 | Photographer: 刘鹰 Ying Liu

项目名称 | Name: 宁静的自由 Liberty in peace (pp.194-201)
项目地址 | Address: 浙江宁波 Ningbo, Zhejiang
建筑面积 | Area: 200 m^2
完成日期 | Completion: 2020.6
主案设计 | Chief Designer: 张奇峰 Qifeng Zhang
软装美陈 | Soft Decor: 袁霞 Xia Yuan
参与设计 | Team Members: 沈璐、梁燕华、叶斌 Lu Shen, Yanhua Liang, Bin Ye
撰文 | Writer: 叶建荣 Jianrong Ye
摄影 | Photographer: 刘鹰 Ying Liu

项目名称 | Name: 共融共生 Harmonious symbiosis (pp.202-207)
项目地址 | Address: 浙江宁波 Ningbo, Zhejiang
建筑面积 | Area:600 m^2
完成日期 | Compeletion: 2018.5
主案设计 | Chief Designer: 卓稣萍 Suping Zhuo
参与设计 | Team Members: 卓永旭、徐群莹、竺财宏 Yongxu Zhuo, Qunying Xu, Caihong Zhu
撰文 | Writer: 叶建荣 Jianrong Ye
摄影 | Photographer: 刘鹰 Ying Liu

项目名称 | Name: 慵懒的家 Home with lazy style (pp.208-219)
项目地址 | Address: 浙江宁波 Ningbo, Zhejiang
建筑面积 | Area: 400 m^2
完成日期 | Completion: 2018.4
主案设计 | Chief Designer: 张奇峰 Qifeng Zhang
软装美陈 | Soft Decor: 袁霞 Xia Yuan
参与设计 | Team Members: 沈璐、梁燕华、叶斌 Lu Shen, Yanhua Liang, Bin Ye
撰文 | Writer: 叶建荣 Jianrong Ye
摄影 | Photographer: 刘鹰 Ying Liu

项目名称 | Name: 空灵维度 Ethereal dimension (pp.220-225)
项目地址 | Address: 浙江宁波 Ningbo, Zhejiang
建筑面积 | Area: 180 m^2
完成日期 | Compeletion: 2019.10
主案设计 | Chief Designer: 刘志明 Zhiming Liu
参与设计 | Team Members: 杨丛丛 Congcong Yang
撰文 | Writer: 刘志明 Zhiming Liu
摄影 | Photographer: 刘鹰 Ying Liu

项目名称 | Name: 设计无痕 Traceless design (pp.226-235)
项目地址 | Address: 浙江宁波 Ningbo, Zhejiang
建筑面积 | Area: 360 m^2
完成日期 | Completion: 2015.6
主案设计 | Chief Designer: 张奇峰 Qifeng Zhang
软装美陈 | Soft Decor: 袁霞 Xia Yuan
参与设计 | Team Members: 沈璐、梁燕华、叶斌 Lu Shen, Yanhua Liang, Bin Ye
撰文 | Writer: 叶建荣 Jianrong Ye
摄影 | Photographer: 刘鹰 Ying Liu

项目名称 | Name: 画境之家 Home with artistic mood (pp.236-245)
项目地址 | Address: 浙江宁波 Ningbo, Zhejiang
建筑面积 | Area: 200 m^2
完成日期 | Completion: 2017.5
主案设计 | Chief Designer: 郑钢 Gang Zheng
撰文 | Writer: 叶建荣 Jianrong Ye
摄影 | Photographer: 张学泉 Xuequan Zhang

T10 设计
T10 Design

联合创始人合影。（从左至右：刘志明、李财赋、卓稣萍（前排）、汪洋、胡梁锋、张奇峰、毛赟、郑钢、潘高峰）
The founders of T10 Design. (Left to right: Zhiming Liu, Caifu Li, Suping Zhuo (front), Yang Wang, Liangfeng Hu, Qifeng Zhang, Yun Mao, Gang Zheng, Gaofeng Pan)

T10设计创立的初心，是希望联合不同领域的杰出设计机构和产业链伙伴，开展技术、空间、生活、美学、艺术的交流和研究，并以跨领域的合作方式和完善的设计产业链服务，为住宅 、地产、办公、商业、会所、酒店、景观等不同类型的客户，提供卓越的设计和专业的服务，创造和谐共生的理想空间与生活方式。

名称中的“T”代表Team(团队)，“10”不仅指10位最初的联合发起人，也代表1和0这两个编织信息世界的基本单位，在设计的世界，也有无限的可能。

T10 Design is a united design team, expecting to communicate with great souls and conducting researches on technology, space, life, aesthetics and art, and wishing to work with partners from other fields to provide complete service in the industry chain of design. They focus on different types of design such as residence, property, offices, shops, clubhouse, hotels and landscape to serve with excellent design skills and professional services, creating harmonious ideal space and lifestyle.

Letter “T” in the name stands for “Team”. “10” not only means ten co-founders, but also represents “1” and “0”, which are two basic units in the information world that can reach infinity.

图书在版编目（CIP）数据

自见：T10设计作品集.2020／T10设计著. —武汉：华中科技大学出版社，2021.4
ISBN 978-7-5680-6976-2

Ⅰ. ①自… Ⅱ. ①T… Ⅲ. ①室内装饰设计-作品集-中国-现代 Ⅳ. ①TU238.2

中国版本图书馆CIP数据核字（2021）第035920号

出　　品：T10设计
主　　编：刘志明
统筹编辑：黄竹盈
英文编辑：高轶超
版式设计：傅行、黄竹盈
官方网址：www.t10design.com

自见：T10设计作品集（2020）
ZIJIAN: T10 SHEJI ZUOPINJI (2020)

T10设计　著

出版发行：华中科技大学出版社（中国·武汉）
武汉市东湖新技术开发区华工科技园
电话：(027)81321913
邮编：430223

策划编辑：王　娜
责任编辑：王　娜
责任校对：赵　萌
责任监印：朱　玢

印　　刷：中华商务联合印刷（广东）有限公司
开　　本：710 mm×1000 mm　1/16
印　　张：16.25
字　　数：156千字
版　　次：2021年4月 第1版 第1次印刷
定　　价：388.00元

華中出版

投稿邮箱：wangn@hustp.com
本书若有印装质量问题，请向出版社营销中心调换
全国免费服务热线：400-6679-118 竭诚为您服务